Security and Privacy From a Legal, Ethical, and Technical Perspective

Edited by Christos Kalloniatis and Carlos Travieso-Gonzalez

Published in London, United Kingdom

IntechOpen

Supporting open minds since 2005

Security and Privacy From a Legal, Ethical, and Technical Perspective
http://dx.doi.org/10.5772/intechopen.78747
Edited by Christos Kalloniatis and Carlos Travieso-Gonzalez

Contributors
Valentina Talimonchik, Ryszard S. Choras, Amit Joshi, Joan Cahill, Katie Crowley, Alison Kay, Sam Cromie, Michael Gormley, Eamonn Kenny, Sonja Hermman, Ciaran Doyle, Ann Hever, Robert Ross, Bata Tripathy, Bhaskar Dasgupta, Nasim Mobasheri, Tanima Chatterjee, Nadine Barrett-Maitland, Jenice Lynch, Ebru Cankaya, Ofer Mendelevitch, Michael Lesh, Christos Kalloniatis, William Stadler, Jaydip Sen, Sidra Mehtab, Aikaterini-Georgia Mavroeidi, Angeliki Kitsiou

First published in London, United Kingdom, 2020 by IntechOpen
IntechOpen is the global imprint of INTECHOPEN LIMITED, registered in England and Wales, registration number: 11086078, 7th floor, 10 Lower Thames Street, London, EC3R 6AF, United Kingdom
Printed in Croatia

British Library Cataloguing-in-Publication Data
A catalogue record for this book is available from the British Library

Additional hard and PDF copies can be obtained from orders@intechopen.com

Security and Privacy From a Legal, Ethical, and Technical Perspective
Edited by Christos Kalloniatis and Carlos Travieso-Gonzalez
p. cm.
Print ISBN 978-1-83881-205-8
Online ISBN 978-1-83881-206-5
eBook (PDF) ISBN 978-1-83881-207-2

For EU product safety concerns:
IN TECH d.o.o., Prolaz Marije Krucifikse Kozulić 3, 51000 Rijeka, Croatia,
info@intechopen.com or visit our website at intechopen.com.

Meet the editors

Dr. Christos Kalloniatis is an Associate Professor in the Department of Cultural Technology and Communication of the University of the Aegean and Director of the Privacy Engineering and Social Informatics (PrivaSI) research laboratory. He is a former member of the board of the Hellenic Authority for Communication Security and Privacy. His main research interests are the elicitation, analysis and modelling of security and privacy requirements in traditional and cloud-based systems, the analysis and modelling of forensic-enabled systems and services, privacy enhancing technologies and the design of information system security and privacy in cultural informatics. He is an author of several refereed papers in international scientific journals and conferences and has served as a visiting professor in many European institutions. He has served as a member of various development and research projects. He lives in Mitilini, the capital of Lesvos island along with his wife Liana and his daughters Elpiniki and Irene.

Carlos M. Travieso-González received his M.Sc. degree in 1997 in Telecommunication Engineering at the Polytechnic University of Catalonia (UPC), Spain; and his Ph.D. degree in 2002 at the University of Las Palmas de Gran Canaria (ULPGC-Spain). He is a Full Professor in Signal Processing and Pattern Recognition and Head of the Signals and Communications Department at ULPGC; teaching from 2001 in subjects on signal processing and learning theory. His research lines are biometrics, biomedical signals and images, data mining, classification systems, signal and image processing, machine learning, and environmental intelligence. He has taken part in 51 international and Spanish research projects, some of them as head researcher. He is co-author of 4 books, co-editor of 24 proceedings books, guest editor for 8 JCR-ISI international journals and up to 24 book chapters. He has had over 440 papers published in international journals and conferences (74 of them indexed on JCR – ISI - Web of Science). He has published 7 patents with the Spanish Patent and Trademark Office. He has been supervisor on 8 PhD theses (12 more are under supervision), and 130 Master theses. He is the founder of The IEEE IWOBI Conference series (and President of its steering committee), The InnoEducaTIC conference series, and The APPIS conference series. He is evaluator of project proposals for the European Union (H2020), Medical Research Council (MRC – UK), Spanish Government (ANECA - Spain), Research National Agency (ANR - France), DAAD (Germany), Argentinian Government, and Colombian Institutions. He has been reviewer in different indexed international journals (<70) and conferences (<220) since 2001. He is a member of IASTED Technical Committee on Image Processing from 2007 and member of IASTED Technical Committee on Artificial Intelligence and Expert Systems from 2011. He will be ACM-APPIS 2021 General Chair and IEEE-IWOBI 2020 and 2020, and was ACM-APPIS 2020 General Chair, IEEE-IWOBI 2019, General Chair APPIS

2019 General Chair, IEEE-IWOBI 2018 General Chair, APPIS 2018 General Chair, InnoEducaTIC 2017 General Chair, IEEE-IWOBI 2017 General Chair, IEEE-IWOBI 2015 General Chair, InnoEducaTIC 2014 General Chair, IEEE-IWOBI 2014 General Chair, IEEE-INES 2013 General Chair, NoLISP 2011 General Chair, JRBP 2012 General Chair and IEEE-ICCST 2005 Co-Chair. He is Associate Editor for the Computational Intelligence and Neuroscience Journal (Hindawi – Q2 JCR-ISI). He was Vice-Dean from 2004 to 2010 at the Higher Technical School of Telecommunication Engineers in ULPGC; and Vice-Dean of Graduate and Postgraduate Studies from March 2013 to November 2017. He won "Catedra Telefonica" Awards in Modality of Knowledge Transfer, in the editions 2017, 2018 and 2019.

Contents

Preface

Understanding and realizing the security and privacy challenges of information systems is a very critical and demanding task for both software engineers and developers to design and implement reliable and trustworthy information systems. These challenges earn even more attention nowadays where our society relies on the use of modern information systems to address its own needs. The number of Internet users that present a high degree of dependability on systems and services for fulfilling their daily activities increases at a great pace. Thus, it is important to understand and identify all the necessary factors that play a key role in satisfying the security and privacy of the system.

This book provides novel contributions and research efforts related to security and privacy by shedding light on the legal, ethical, and technical aspects of security and privacy. This book consists of 12 chapters divided in three groups. The first contains works that discuss the ethical and legal aspects of security and privacy, the second contains works that focus more on the technical aspects of security and privacy, and the third contains works that show the applicability of various solutions in the aforementioned fields.

This book is perfect for both experienced readers and young researchers that seek to study the different sub-disciplines of security and privacy. All authors that contributed in this book presented their works in a clear and easy to read way in order for everyone to be able to understand the meaning of their contribution in relation to the book's objectives. Since every author has a background in security and privacy and introduces new research topics and solutions through their work, this book can serve also as a valuable handbook for teaching purposes.

I would like to thank all authors for their contributions and the editorial team of IntechOpen for their valuable support to finish and publish this book. I would also like to thank my wife Liana for her support during the preparation of this book.

Christos Kalloniatis
Associate Professor,
University of the Aegean,
Greece

Carlos Travieso-Gonzalez
University of Las Palmas de Gran Canaria,
Spain

Section 1

Legal and Ethical Issues

Chapter 1

Legal Aspects of International Information Security

Valentina Petrovna Talimonchik

Abstract

The objective of the research is considering the international information security concept that has developed at the global and regional levels and analysis of legal instruments for its implementation and resolving problems of the regulation of relations in the global information society. A complex of general scientific and philosophical methods including the formal-logical, comparative-legal, formal-legal, systemic-structural, and problematic-theoretical methods, as well as methods of analysis and synthesis, generalization and description, and comparison, was used in the research. As a result of the research, it has been found that a unified concept of provision of the international information security has developed at the global and regional levels, which needs legal instruments for its implementation at the global level. In the drafting and acceptance of international treaties at the global level, the experience of the Council of Europe in prosecution of cybercrime and protection of privacy should be used. The findings can be used in the activities of international organizations in execution of their functions of unification and harmonization of the international information security law and by the national telecommunication operators in the process of entering international and foreign markets.

Keywords: international law, global level, regional level, information security, cyberterrorism, computer crimes, privacy

1. Introduction

The technological progress has led to radical changes in the contemporary world. The system of international relations changed. The development of information and communication technologies (ICT) has affected all the areas of public life including the economy, politics, social issues, and culture, bringing them together in the framework of establishment of an information society.

By the present time, the information society concept has been represented in a number of international documents among which are the Declaration of Principles entitled "Building the Information Society: a Global Challenge in the New Millennium" (hereinafter referred to as the 2003 Declaration) and the Plan of Action of the World Summit on the Information Society of December 12, 2003.

Information society is a more general category as compared to the global information society. It can be established within a single state or at the regional or global levels. At the global level, it will be referred to as the global information society.

The global information society can be defined as a system of international relations that are established in the sphere of operation of information systems, which are based on information and communication technologies, in which international information relations affect political, economic, social, and cultural relations. At the same time, the states participate in relations in the global information society as equal subjects of international information relations.

The development of ICT is related to the effect on established branches and institutes of international law as well as to the regulation of new relations that arise as a result of ICT development.

The most complicated problem is the effect of ICT on established branches and institutes of international law. The mechanism for the development of international law provisions is such that legal regulations tend to "fall behind" the level of ICT development.

Currently, the spreading and use of ICT affect the interests of the entire international community; these technologies can potentially be used for purposes that are incompatible with the objectives of international stability and security and can have an adverse effect on the integrity of the infrastructure of the states, disturbing their security in the civil and military areas.

The efforts of individual states are insufficient for ensuring international information security. First of all, the prohibition on the use of information weapons by states must be established in international law. Separate regulation is required for matters of information security of individuals (protection from defamation and privacy).

The forming special principles of international information law include the principle of confidentiality and security in using ICT. Strengthening the trust framework, including information security and network security, authentication, privacy, and consumer protection, is a prerequisite for the development of the information society and for building confidence among users of ICTs. A global culture of cyber security needs to be promoted, developed, and implemented in cooperation with all stakeholders and international expert bodies. These efforts should be supported by increased international cooperation. Within this global culture of cyber security, it is important to enhance security and to ensure the protection of data and privacy while enhancing access and trade. In the 2003 Declaration, the term "cyber security" has a wider meaning that only protection from cybercrimes. In particular, the Declaration notes that the summit participants support activities of the United Nations to prevent the potential use of ICTs for purposes that are inconsistent with the objectives of maintaining international stability and security and may adversely affect the integrity of the infrastructure within states, to the detriment of their security.

These regulations ensure the relation of the developing principle of international information law with the existing principles, namely, the principle that the exercise of freedom of opinion, expression, and information is an essential factor in strengthening peace and international security; the principle that the media should contribute to the strengthening of peace and international understanding and to the struggle against racism, apartheid, and incitement to war; and the principle of the need to publicize the denunciation of information, the spreading of which has caused damage to efforts of strengthening of peace and international understanding, the development of human rights, and the struggle against racism, apartheid, and incitement to war.

The problems of information security of individuals and legal entities have been examined in fundamental research on the comparative law of information technologies by Bainbridge [1], Campbell [2], Rowland and Macdonald [3], Smedinghoff [4], and Black [5].

The issue of privacy protection, primarily using national legal instruments, has been covered in particular chapters in the fundamental research on the law of information technologies by Bell and Ray [6], Reed [7], and Angel [8] and special research by Solove [9] and Nouwt, Berend, and Prins [10].

Technical and organizational aspects of ensuring information security have been covered in the works of Egan and Mather [11], Hunter [12], and Volonino and Robinson [13].

The matter of implementation of the concept of ensuring international information security has already been considered in research, although the concept itself has not been stipulated. Lloyd [14] considered the acts of the UN, the Council of Europe, OECD, and the Asia-Pacific Community when addressing the issues of privacy, primarily considering "soft law" acts. In a review of cybercrime problems, this author gives a brief overview of the Council of Europe Convention on Cybercrime, the OECD Guidelines for the Security of Information Systems, and the EU acts.

The contents and significance of the Convention on Cybercrime of November 23, 2001, have been discussed in the studies by Lloyd [14], Murray [15], and Koops, Lips, Prins, and Schellekens [16]. But these studies did not cover the problems of using the experience of the Council of Europe at the global level.

With regard to the 2001 Convention, Hopkins [17] has noted its excessive broadness and lack of clarity in its basic terms. For example, this Convention defines a computer system as any device or a group of interconnected or related devices, one or more of which, pursuant to a program, performs automatic processing of data. In such case, the term device will include children's toys, Palm Pilots, and cable television devices. Therefore, the scope of the 2001 Convention extends from real computer crimes to interference in any devices where software is used.

The concept of personal data in international acts has been criticized in the legal doctrine. In particular, Berčič and George [18] state that this definition is too broad because any information about a person can be regarded as personal data (e.g., information that an individual is wearing a red shirt). On the other hand, there arise practical complicacies with attributing certain data as personal data (e.g., social security identification numbers).

Polcak [19] has pointed out that in various European countries, there are complicacies with attributing IP addresses, personal telephone numbers, data entered anonymously when receiving services via the Internet, and data of deceased persons as personal data.

The absence of unified list of personal data in the national legal systems is the reason of the imperfection of the international legal regulation. The efforts made in the area of harmonization have not been successful enough. This is confirmed by the attempts that are being made at the national level to create an own definition of personal data. In particular, a number of authors have named the Durant v. FSA case in British courts as an example. In this case, the Court of Appeals has defined personal data as information that affects the privacy of the data subject including their personal and family life and business or professional abilities [20].

It should be noted that currently, proposals to make global international treaties primarily come from non-state actors. In August 2000, a group of researchers from Stanford University presented the Draft International Convention to Enhance Protection from Cyber Crime and Terrorism (the Stanford Project). Brown drafted a convention regulating the use of information systems in armed conflicts. On November 6, 2009, the International Conference of Data Protection and Privacy Commissioners adopted a resolution entitled "Standards of Privacy and Personal Data," for which it established a work group to develop a draft global treaty and listed the criteria for the drafting of it. It is planned to submit the developed sections of the treaty to the UN. Thus, researchers and international forums are proposing specific projects, but no systemic work is carried out in the framework of the UN, International Telecommunication Union (ITU), or UNESCO.

At the same time, there are no monographic researches of the general concept of international information security that would cover the regional and global levels and the problems of development of its legal basis.

The present study, based on the analysis of international acts, reveals the content of the general concept of international information security that would cover the regional and global levels. "Soft law" acts are appropriate for the formulation of the general concept of international information security, but not for its implementation. Therefore, the author proposes a draft convention with the purpose of creating of global network of information security.

2. Analysis of international acts

The objective of the research is consideration of the international information security concept that has developed at the global and regional levels and formulation proposal for elaboration of legal instruments for its implementation in connection with the concept of the global information society. For this, the analysis of existing international information security system at the global and regional levels shall be made, a description and a generalization of the analysis results. For the analysis of existing international information security system, formal-logical, systemic-structural, and problematic-theoretical methods have been used. At the same time, comparative-legal method is used to analyze the provisions of information security at the global and regional levels.

In order to solve the problems of international security that have arisen with the development of ICT, the UN General Assembly has adopted resolutions entitled "Developments in the field of information and communications in the context of international security" at each of its sessions since 1998. The main idea of these resolutions is that the significant progress, which has been achieved in the development and implementation of the latest information technologies and telecommunications, has caused negative consequences as well as positive ones. At the same time, the positive consequences, namely, new opportunities for the entire mankind, are obvious.

However, the UN General Assembly has expressed concern that new technologies and facilities that these technologies and means can potentially be used for purposes that are inconsistent with the objectives of maintaining international stability and security and may adversely affect the integrity of the infrastructure of states to the detriment of their security in both civil and military fields.

The resolutions invite states to inform the UN Secretary-General on the following issues, namely, (1) general assessment of the problems of information security, (2) development of concepts relating to information security, and (3) development of international principles aimed at ensuring information security of global information and telecommunications systems and combating information terrorism and crime.

It should be noted that there exist resolutions which confirm a certain progress in ensuring information security. They contain specific proposals for the development of an information security system that can be used for the draft of relevant international treaties. For example, the UN General Assembly adopted the Resolution No. 58/199 of December 23, 2003, on the creation of a global culture of cybersecurity and the protection of critical information structures, which defines elements for protection of critical information infrastructures, namely, (1) having emergency warning networks regarding cyber-vulnerabilities, threats, and incidents; (2) raising awareness to facilitate stakeholders' understanding of the nature and extent of their critical information infrastructures and the role each must play in protecting them; (3) examining infrastructures and identifying interdependencies among them, thereby enhancing the protection of such infrastructures; and (4) promoting partnerships among stakeholders, both public and private, to share and analyze critical infrastructure information in order to prevent, investigate, and respond to damage to or attacks on such infrastructures, etc.

The nature of the elements for protection of the most important information structures is such that they can be included in an international treaty if they are specified.

Currently, an institutional mechanism for ensuring international information security has been established in the framework of the UN. States submit their assessments of the condition of information security on a regular basis, which are included in the reports of the Secretary-General and have contributed to a better understanding of the essence of problems of international information security and related concepts.

The work of the Group of Governmental Experts on Developments in the Field of Information and Telecommunications in the Context of International Security and the resulting report (2015) have been quite effective. The Group concluded that international law and, in particular, the Charter of the United Nations are relevant and important for the maintenance of peace and stability and the development of an open, safe, stable, accessible, and peaceful information environment; that voluntary and non-binding standards, rules, and principles of responsible behavior of states in the use of information and communication technologies can mitigate the risk of violation of international peace, security, and stability; and that, subject to the unique features of the information and communication technologies, more standards can be developed over time.

In addition, the EU, OAS, and Caribbean Community (CARICOM) have achieved certain results in the development of regional concepts of the improvement of information security. For example, on February 7, 2013, the Joint Communication to the European Parliament, the Council, the European Economic and Social Committee, and the Committee of the Regions entitled "Cybersecurity Strategy of the European Union: An Open, Safe and Secure Cyberspace" was adopted. The strategy contains principles for cyber security, strategic priorities, and actions. The principles of cybersecurity include the principle that the EU's core values apply as much in the digital as in the physical world; protecting fundamental rights, freedom of expression, personal data and privacy; access for all; democratic and efficient multi-stakeholder governance; and a shared responsibility to ensure security.

In order to support member states in their fight against cybercrime, OAS, through the Inter-American Committee Against Terrorism (CICTE) and the Cyber Security Program, is committed to developing and furthering the cyber security agenda in the Americas. Cooperating with a wide range of national and regional entities from the public and private sectors on both policy and technical issues, the OAS seeks to build and strengthen cyber security capacity in the member states through technical assistance and training, policy roundtables, crisis management exercises, and exchange of best practices related to information and communication technologies.

CARICOM Ministers with responsibility for information and communication technologies met on May 19, 2017, as efforts continue to move on the establishment of the CARICOM Single ICT Space. Several preparatory meetings of officials were held to advance work on the Integrated Work Plan for the Single ICT Space and the draft Terms of Reference for the CARICOM-US Joint Task Force. The Integrated Work Plan will set out the activities that need to be completed for the development of the Single ICT Space. The activities of the work plan will focus on areas such as conducting gap analyses, public awareness, specific telecommunications issues, legal and regulatory reform for cyber security, bringing technology to the people, resource mobilization, as well as forecasting for the CARICOM Digital Agenda 2025. The Single ICT space and the Region's Digital Agenda 2025 will be constructed on the foundation of the Regional Digital Development Strategy (RDDS) which was approved in 2013 and will also have inputs from the Commission on the Economy and the Post-2015 Agenda.

The concept of international information security is developing in the framework of soft law. International treaties in this field are quite scarce.

The privacy problem has been represented in the international law. Currently, the privacy provision is contained in many international documents. Of particular importance is Article 12 of the 1948 Universal Declaration of Human Rights, which stipulates that no one shall be subjected to arbitrary interference with his privacy, family, home, or correspondence nor to attacks upon his honor and reputation. Everyone has the right to the protection of the law against such interference or attacks. States recognize noninterference in personal and family life as a fundamental human right. It should be noted that the 1948 Universal Declaration is a recommendatory act, but a number of its provisions represent the established international customs. At the same time, the right to protection of private life may be restricted, which makes it impossible to regard it as a right that is recognized unconditionally.

Currently, the protection of privacy has a treaty origin. Provisions for protection of privacy are stipulated in Article 17 of the 1966 International Covenant on Civil and Political Rights, Article 8 of the 1950 European Convention for the Protection of Human Rights and Fundamental Freedoms, and Article 11 of the 1969 American Convention on Human Rights.

Article 12 of the 1948 Universal Declaration of Human Rights has been incorporated into Article 17 of the 1966 International Covenant on Civil and Political Rights. Everyone has the right to the protection of the law against such interference or attacks. Similar provisions are stipulated by regional international treaties.

It appears quite reasonable to abolish the unification of the concept of privacy and personal data as a component of privacy in international law. Privacy is an area where individual needs of a person to be left to himself/herself are revealed. Every individual will delineate the limits of his/her privacy to himself/herself. Contemporary international law is limited to the regulation of matters of collection, processing, storage, and transfer of personal data, which are not the only issues of privacy. It appears that the privacy provision in the International Covenant on Civil and Political Rights is quite generalized but does not require specification in the information age, as it enables any individual to protect privacy in every case when the individual so wishes.

The problem of personal data protection in the framework of information security problems is perfectly reasonable to be considered. Information security is a category applicable to all subjects of information relations including states and non-state (legal entities, individuals, TNCs, nongovernmental organizations, etc.) ones. Information security of individuals is related to the respect of their privacy in the information sphere, protection from defamation, libel, insults, psychological pressure, information terrorism, etc. Therefore, the legal problems of privacy in the information sphere are a component of legal regulation of information security of the individual.

If one tries to define the content of privacy in the information area, it will be different for every individual. In the information sphere, the range of data that a person tries to make inaccessible to the public is always different. For example, one person will not hide the fact that they are infected with HIV and may say it in an interview to a journalist, while another person will choose to not even tell close friends about it. Thus, the boundaries of privacy are always individual.

Contemporary international law provides limited privacy protection because it cannot adapt to the needs of each individual due to the general nature of the provisions. At the same time, the current international acts do not contain a list of personal data but give a fairly wide definition of such data.

An identical approach to the definition of personal data is characteristic of the OECD Guidelines governing the Protection of Privacy and Transborder Flows of Personal Data of September 23, 1980, and the 1981 Convention for the Protection of Individuals with regard to the Automatic Processing of Personal Data. In these documents, personal data are defined as any information relating to an identified or identifiable individual. Therefore, protected data include any information about an individual that can be identified. Such a broad range of protected information makes it possible to protect personal data in the situation of changing technologies that are used to collect and process data. In particular, protected data include PIN codes, logins, passwords, etc.

Despite the quite broad definition of personal data in international documents, the concept of personal data is somewhat narrower than privacy in the information area. Based on the provision of the Universal Declaration, the concept of privacy includes not just personal but also family secrets as well as the secret of correspondence. Personal data only relate to data about identified or identifiable individual. Certain provisions are applied only to individual, information on whom is stored in a particular system. For example, the 1981 Convention stipulates that any individual has the right to establish the existence of an automated personal data file, its main purposes, as well as the identity and habitual residence or principal place of business of the controller of the file; to obtain at reasonable intervals and without excessive delay or expense confirmation of whether personal data relating to him are stored in the automated data file as well as communication to him of such data in an intelligible form; to obtain, as the case may be, rectification or erasure of such data if these have been processed contrary to the provisions of domestic law giving effect to the basic principles set out in Articles 5 and 6 of this Convention; etc.

Therefore, the right to access, correct, and destroy personal data is recognized only for the person whose data have been collected. However, family secret is a different term. For example, one may conceal data about a disease of one's child or husband or addictions of deceased relatives. In essence, while personal data relate to one person, family secret is kept in a certain family and affects its collective private interests. Disclosure of family secret can harm both individual and the family as a whole including family breakdown and ruined relationships.

The existing special international acts that protect personal data in the course of their automated processing contribute to protection of not just personal but also family secrets. However, they offer no direct protection of family secrets.

As for the confidentiality of correspondence, certain provisions for telecommunications are contained in the Convention of the International Telecommunication Union. Article 40 of the ITU Convention provides for the secrecy of telecommunication messages. Government telegrams and service telegrams may be expressed in secret language in all relations. Private telegrams in secret language may be admitted between all Member States with the exception of those which have previously notified, through the Secretary-General, that they do not admit this language for that category of correspondence. Member States which do not admit private telegrams in secret language originating in or destined for their own territory must let them pass in transit, except the Constitution. ITU does not have the power to regulate information on the Internet including measures for ensuring its confidentiality. At the regional level, a provision on the confidentiality of electronic communications is stipulated at the EU. The relevant provision is contained in the Directive 2002/58/EC of the European Parliament and of the Council of July 12, 2002, concerning the processing of personal data and the protection of privacy in the electronic communications sector.

The most progressive in privacy protection is the EU experience. This integration organization has adopted the Regulation No. 2016/679 of the European Parliament and of the Council on the protection of natural persons with regard to the processing of personal data and on the free movement of such data, and repealing the Directive 95/46/EC (the General Data Protection Regulation) of April 27, 2016. This act is of direct effect and application in the EU Member States. A feature of the General Regulation is that any processing of personal data in the context of the activity of establishing a controller or data processing entity in the Union must be performed in accordance with the Regulation regardless of whether the data processing is affected within the Union. In order to ensure that individuals are not deprived of the protection provided by the Regulation, processing of personal data of data subjects located in the Union by a controller or data processing entity that have not been established in the Union must be governed by this Regulation if the data processing relates to the supply of goods or services to such data subjects regardless of payment. The Regulation establishes a certain legal regime for personal data processing including the conditions for their processing and requirements to their storage and transfer. The processing of personal data by public authorities, computer emergency response teams (CERTs), computer security incident response teams (CSIRTs), providers of electronic communication networks and services, and providers of security technologies and services is a legitimate interest of the relevant data controller to the extent that it is necessary and adequate as compared to the objectives of providing network and information security, i.e., the ability of the network or information system to resist (with a given level of confidence) accidental events and illegal or intentional acts that compromise the availability, authenticity, integrity, and confidentiality of stored or transferred personal data as well as the safety of the relevant services transferred via such networks and systems. Protection of privacy within the EU is also supported by the EU Court. In the Maximillian Schrems v. Data Protection Commissioner case (complaint No. C362/14), the transfer of personal data by Facebook in the USA was appealed against in the framework of the Principles of Privacy program. The EU Court concluded that the Commission had not stated in its Resolution that the USA had actually provided an adequate level of protection by virtue of their laws or international obligations. Therefore, without having to examine the content of the Principles of Privacy, Resolution 2000/520 did not comply with the EU acts in the field of privacy and is therefore invalid.

However, the EU experience takes account of the patterns of functioning of integration organizations and requires significant adaptation for use at the global level.

At the regional level, two conventions have been adopted where computer crimes are regarded as crimes of international nature. These are the Convention on Cybercrime of November 23, 2001 (hereinafter referred to as the 2001 Convention) and the Commonwealth of Independent States Agreement on Cooperation in Combating Offenses related to Computer Information of June 1, 2001 (hereinafter referred to as the CIS Agreement).

The basic ideas of these conventions are the definition of unified elements of computer crimes, which the states should include in their national law, and development of measures for combating such crimes.

The CIS agreement has no definition of a computer system whatsoever, which results in an uncertainty with regard to the object of infringement.

Both the 2001 Convention and the CIS Agreement contain definitions of computer data. However, the definition in the Agreement is more concise; namely, it is information stored in computer memory, on machine or other device, in a form that is accessible to perception or transfer via communication channels. This definition is incomplete.

The 2001 Convention offers a broader concept; namely, computer data includes any representation of facts, information, or concepts in a form suitable for processing in a computer system, including a program suitable to cause a computer system to perform a function. As a result, the CIS Agreement does not cover any software that is inaccessible to human perception but causes computer systems to operate. Interference in such software is dangerous for the public. In this case, the broader approach in the 2001 Convention should be considered justified.

The CIS Agreement contains an attempt to define computer crime, which cannot be regarded as successful. A crime in the field of computer information is described as a criminal offense, the object of infringement in which is computer information. This definition is different from the definition that has been accepted in the doctrine. It is not mentioned that computer information can be both the object and the means of an offense.

The 2001 Convention contains a number of terms that are unknown to the CIS Agreement, namely, *service provider* and *data flows*. The need to use these terms is due to the fact that the 2001 Convention defines a broader range of measures for combating computer crime than the CIS Agreement.

As for standardized elements of computer crimes, they are different in the 2001 Convention and the CIS Agreement. Some crimes have the same title but different meanings. For example, the 2001 Convention and the CIS Agreement state that illegal access to information is a criminal offense. However, the CIS Agreement is very laconic. It regards illegal access to information that is protected by law as a criminal offense if such act has caused destruction, blocking, modification or copying of information, or disruption of the operation of computers, computer systems, or their networks. The 2001 Convention stipulates that illegal access to a computer system as a whole or a part of it is a crime by itself, without stating any extra qualifying features. Therefore, the 2001 Convention prosecutes any illegal access to computer systems, while the CIS Agreement is limited to access that has led to certain consequences.

The 2001 Convention includes a number of crimes that are not covered by the CIS Agreement. These are illegal data interception, data and system interference, misuse of devices, computer-related forgery, computer-related fraud, and crimes related to child pornography. A special feature of the 2001 Convention is that it covers certain common crimes (forgery, fraud) which become much more dangerous because they are committed using computers.

Therefore, the CIS Agreement uses a narrower approach to the concept of computer crime. These are only the crimes that infringe on the security of computer systems, i.e., the protected object is computer systems as such. The 2001 Convention criminalizes a broader range of acts where computer systems can be the object of or the means for committing the offense. The approach to the definition of computer crime in the 2001 Convention is more correct.

The existing contradictions in the content of international treaties on combating computer crime may result in difficulties for the states that are parties to both treaties. Basically, the provisions of the two treaties are mutually exclusive, which complicates their simultaneous application.

It should be noted that the 2001 Convention contains references to a number of international treaties. The issues of the relationship between the 2001 Convention and the CIS Agreement shall be resolved with consideration of clause 2 of Article 39 of the 2001 Convention. If two or more parties have already concluded an agreement or treaty on the matters dealt within this Convention or have otherwise established their relations on such matters, or should they in future do so, they shall also be entitled to apply that agreement or treaty or to regulate those relations accordingly. However, where parties establish their relations in respect of the matters dealt within the Convention other than as regulated therein, they shall do so in a manner that is not inconsistent with the Convention's objectives and principles.

Therefore, in the case if a state is a party to both of the abovementioned international treaties, the CIS Agreement will apply to the same matter.

Article 13 of the CIS Agreement stipulates that this agreement does not affect the rights and obligations of the parties arising out of other international treaties to which they are parties. Therefore, it allows the application of the 2001 Convention.

The existence of various regulations regarding their correlation in the considered international treaties suggests that their practical application may be complicated. For example, the states may experience difficulties in choosing the legal aid procedure. Such issues should be resolved by consultations between the states concerned.

However, in view of the harmonization nature of international treaties and the fact that the content of the 2001 Convention is broader, in the case if a state is a party to the two treaties at the same time, such state shall implement the 2001 Convention and, in the part where the provisions of the treaties are different, the CIS Agreement, as this is allowed by the 2001 Convention itself.

3. Proposal for global network of information security

The concept of developing a comprehensive system of international security is useful because of its systemic nature. This concept is not limited to military security issues but also covers economic, political, humanitarian, and information security. It should be noted that this concept needs to be clarified. Since it concerns the development of a comprehensive system of international security, it should cover the entire system of international relations. The concept of developing a comprehensive system of international security also applies to non-state international relations.

A comprehensive system of international security means a status where the interstate system is protected from the dangers that exist in contemporary world. It implies stable functioning of the system of international relations. Relations between subjects of the interstate system also include information relations. The system of such relations includes interstate and non-state relations.

Information security should be considered in two aspects.

If the systemic approach is applied, information security will act as a backbone element. It can be regarded as a status of the international relation system, which is described by stability and security from information weapons and threats.

In addition, information security can be regarded as an ideal model. There are conceptual ideas what exactly information security should be like. It is regarded in the sociological (as a certain state of social relations), technical (compliance with standards and other technical requirements), and legal (compliance with prohibitions and restrictions on the spreading of data) aspects. Based on conceptual ideas, information security can be defined as a model for stable functioning of the information relation system.

The comprehensive system of international security and information security has a certain sphere of intersection. Information security of the international system is a component of the comprehensive system of international security. However, international relations are more than just relations between subjects of international public law. The requirement of information security is equally applicable to international non-state and domestic relations.

When one uses the concept of international information security, one may define this concept based on the more general concept of information security. If one distinguishes between domestic and international information security, the first one relates to domestic information relations and the second one, to international information relations. In each of the systems of relations, information security has common features; namely, it serves as a backbone element and ensures a stable state of the system of information relations. Therefore, international information security is a status of the international information relation system, which is described by stability and security from information weapons and threats.

The development of the concept of international information security has led to the appearance of terms in the legal doctrine that had not previously been known in the practice of states. Currently, researchers use terms such as information weapons, information terrorism or cyberterrorism, and information crime or cybercrime.

The state of international legal regulation is such that these new terms have not been stipulated in treaties (save for computer crimes). However, a number of social phenomena evidence that these terms should be regarded as destabilizing factors for the system of international relations.

As for information weapons, they can be described quite generally as any means of affecting the mass and individual consciousness, which can damage, distort, destroy, or conceal data. A special feature of information weapons is that they are not used in the military field alone. Information weapons can be used for committing computer crimes, hacker attacks causing property damage, etc. The use of information weapons has been known in international practice since the second half of the twentieth century. For example, it was used widely in the Palestinian-Israeli conflict.

With the adoption of individual conventions on cybercrime, there appeared a trend in international law to prosecute the consequences of the use of information weapons rather than the weapons as such.

It should be noted that the use of information weapons has various scales. For example, information terrorism can be regarded as one of the most dangerous use of information weapons.

Information terrorism can be defined as using information weapons for undermining the constitutional order of other states or the international legal order and international relations in general.

Cyberterrorism comprises both direct terrorist activities with the use of computers, networks and data in networks, and various supplementary operations including coordination, preparation, and organization of terrorist activities using networks and data in networks and spreading knowledge about terrorism and terrorists' skills.

Individual examples of cyberterrorism have been known from the second half of the twentieth century. In 1985, a radical leftist group in Japan attacked the united railway management network using computer systems. Fortunately, the computers of the railway had good protection, which could not be hacked.

The 2001 Convention takes no account of the special features of cyberterrorism. It only takes account of "ordinary" crimes.

In this paper, computer crime is understood in the broad sense as any crime committed by using computer networks, software, or individual computers.

However, in the international law, the term *computer crime* will always have a special meaning, which is not necessarily the same as the meaning of this term in the national law. Some crimes that are punishable under the laws of one state do not affect the interests of another state or the international community as a whole.

While international crimes are threatening for the international peace and security, crimes of an international nature are common crimes in combating which states cooperate.

International crimes can be committed using computers. Global computer networks enable propaganda of war, genocide, apartheid, and racial discrimination. Moreover, the use of computers for military technology can lead to electronic communications becoming a means of aggression.

It should be noted that the existing international treaties on computer crime regard computer crimes primarily as crimes of an international nature. They define the elements of crimes that must be criminalized in national law as well as measures of international cooperation in combating such crimes.

The development of legal foundations of the global information society is to a great extent spontaneous. In the framework of the institutional mechanism of cooperation between states, there is not enough systemic vision of what the legal regulation should be like to meet the development of the technological progress.

Therefore, the information society concept needs a corresponding integral concept of international legal regulation of information exchange relations in the information society.

Some objectives have existed for a long time and are related to a lack of regulation of certain problems (matters of combating computer crime, protection of privacy at the global level, etc.), while others have appeared relatively recently as a result of technological progress.

What are the objectives that should be addressed at the global level? When determining the range of objectives, one should consider that information technologies have become global and reveal the interdependence of the contemporary world. At the same time, there is the experience of regulation of electronic data exchange relations in the framework of the Council of Europe, which should be recognized as progressive and useful for the global level.

The primary objective for the global level is solving the problems that have already been solved in the framework of the Council of Europe (combating computer crime, protection of personal data). The models of the Council of Europe have already been tested in practice, and in any case they have no significant alternative.

For the prosecution of computer crimes and protection of privacy, the global network of international information security can be created under the Security Council of UN decision by adoption of the international treaty. The global network of international information security shall provide for search in computer networks performed in one state on request of another state, real-time collection of traffic data and real-time collection and interception of content data. Therefore, the general mechanism of legal aid shall be applied, but its content is special.

In the global network of international information security, any state may request another state to order or otherwise obtain the expeditious preservation of data stored by means of a computer system, located within the territory of that other state.

The global network of international information security stipulates 24–7 access, i.e., each state shall designate a contacting board available on a 24-hour, 7-day-a-week basis, in order to ensure the provision of immediate assistance for the purpose of investigations or proceedings concerning criminal offenses related to computer systems and data or for the collection of evidence in electronic form of a criminal offense. Basically, this procedure can take a few minutes.

The global network of international information security can also provide the access for non-state actors in privacy violations and defamation cases.

One state may get access to publicly available computer data, regardless of their geographical location, without permission of any other state. This primarily applies to data that are contained on the Internet. If a website has no access codes and the data on it can be accessible to everyone, it can be used by search, investigative, and judicial authorities. It is a general practice of access to data in open computer networks. There exists an international custom, according to which states do not put special restrictions on the spreading of publicly available data in computer networks. Special regulations are established for data, the spreading of which is prohibited or restricted. If any person may have access to information, it would be illogical to deny such access to law enforcement authorities.

In addition, any state can access, through a computer system in its territory, stored computer data, if the state obtains the lawful and voluntary consent of the person or legal entity who has the lawful authority to disclose the data. In this case, the state body of one state must address the provider, which is located in another state, directly.

Therefore, the development of the institute of mutual legal assistance in criminal matters, which is affected by the struggle with computer crimes, is not just about introducing electronic communication technologies in traditional types of legal assistance and not just about specifying legal aid measures in relation to electronic communication technologies but also about radical change in the very content of this institute.

The system of international information security is establishing at the moment. The international information security of the interstate system is a component of the comprehensive system of international security. At the same time, international information security is a stabilizing factor in the system of non-state international relations. However, a number of threats to international information security affect the field of both interstate and international non-state relations. In "soft law" acts, a unified concept of the development of a system of international information security has been elaborated at the global and regional levels. However, "soft law" acts are not suitable for its implementation. They can contribute to development of international customs, but that can take a considerable time. Therefore, global international treaties should be drafted.

The 2001 Convention and the CIS Agreement have become the first international treaties that stipulate a system of measures for combating a specific type of crime in the field of information, namely, computer crimes. Formerly, information crimes had been covered in particular international treaties along with other crimes (such as propaganda of racial discrimination). The treaties considered have a very important role, as they have established the foundations of the jurisdiction of states for criminal cases on the Internet and the rules of international cooperation that ensure coordinated actions of states in combating computer crimes. Despite some shortcomings of the treaties, as a whole they provide for systems of interrelated international and national measures for combating computer crimes and can be the basis for drafting of a global international treaty.

Author details

Valentina Petrovna Talimonchik
Saint Petersburg State University, Russia

*Address all correspondence to: talim2008@yandex.ru

References

[1] Bainbridge DI. Introduction to Information Technology Law. Edinburg: Pearson Education Limited; 2008

[2] Campbell D, Ban C, editors. Legal Issues in the Global Information Society. New York: Oceana Publications Inc; 2005

[3] Rowland D, Macdonald E. Information Technology Law. Abingdon: Cavendish Publishing Ltd; 2005

[4] Smedinghoff TJ, editor. Online Law. New York: Pearson Education Corporation; 2000

[5] Black SK. Telecommunications Law in the Internet Age. San Francisco: Morgan Kaufmann Publishers; 2002

[6] Bell R, Ray NEU. Electronic Communications Law. Richmond: Richmond Law and Tax Ltd; 2004

[7] Reed C. Internet Law: Text and Materials. Cambridge: Cambridge University Press; 2005

[8] Reed C, Angel J, editors. Computer Law: Law and Regulation of Information Technology. Oxford: Oxford University Press; 2007

[9] Solove DJ. The Digital Person: Technology and Privacy in the Information Age. New York: New York University Press; 2004

[10] Nouwt S, Berend R. In: Prins V, editor. Reasonable Expectations of Privacy? The Hague: ITeR; 2005

[11] Egan M, Mather T. The Executive Guide to Information Security: Threats, Challenges and Solutions. Indianapolis: Addison-Wesley; 2005

[12] Hunter J. An Information Security Handbook. London: Springer Verlag London Limited; 2001

[13] Volonino L, Robinson SR. Principles and Practice of Information Security. Upper Suddle River: Pearson Education Inc; 2004

[14] Lloyd IJ. Information Technology Law. Oxford: Oxford University Press; 2008

[15] Murray A. Information Technology Law: Law and Society. Oxford: Oxford university press; 2010

[16] Koops BJ, Lips M, Prins C, Schellekens M. Starting Points for ICT Regulation. The Hague: T.M.C. Asser Press; 2006

[17] Hopkins S. The Cybercrime Convention Does Not Provide Substantive Lawmaking Guidance. 2018. Available from: http://www.netdialogue.org/discussion/?p=23 [Accessed: 12 March 2018]

[18] Berčič B, George C. Identifying personal data using relational database design principles. International Journal of Law and Information Technology 2009. V. 17. N 3. P. 234-235

[19] Polcak R. Aims, methods and achievements in European data protection. International Review of Law, Computers & Technology. 2009. V. 23. N 3. P. 183

[20] McCullagh K. Protecting "privacy" through control of "personal" data processing: A flawed approach. International Review of Law, Computers & Technology. 2009. V. 23. N 1-2

Chapter 2

Ethical Issues in the New Digital Era: The Case of Assisting Driving

Joan Cahill, Katie Crowley, Sam Cromie, Alison Kay, Michael Gormley, Eamonn Kenny, Sonja Hermman, Ciaran Doyle, Ann Hever and Robert Ross

Abstract

Mobility is associated with driving a vehicle. Age-related declines in the abilities of older persons present certain obstacles to safe driving. The negative effects of driving cessation on older adults' physical, mental, cognitive, and social functioning are well reported. Automated driving solutions represent a potential solution to promoting driver persistence and the management of fitness to drive issues in older adults. Technology innovation influences societal values and raises ethical questions. The advancement of new driving solutions raises overarching questions in relation to the values of society and how we design technology (a) to promote positive values around ageing, (b) to enhance ageing experience, (c) to protect human rights, (d) to ensure human benefit and (e) to prioritise human well-being. To this end, this chapter reviews the relevant ethical considerations in relation to assisted driving solutions. Further, it presents a new ethically aligned system concept for assisted driving. It is argued that human benefit, well-being and respect for human identity and rights are important goals for new automated driving technologies. Enabling driver persistence is an issue for all of society and not just older adult.

Keywords: driverless cars, older adults, ethics, well-being, self-efficacy

1. Introduction

Mobility is defined as "the ability to move oneself (either independently or using assistive device or transportation) within environments that expand from one's home to the neighbourhood and regions beyond" [1]. The ability to move about the community is essential for carrying out the instrumental activities of daily living (i.e. basic life-maintenance activities) and ensuring social participation [1].

Growth in ageing populations is a global trend. A recent United Nations report states that the number of persons aged 60 (or older) is expected to grow from 962 million in 2017, to 2.1 billion in 2050, and 3.1 billion in [2]. According to the Global Status Report on Road Safety published by The World Health Organization (WHO), approximately 1.35 million people around the world die each year in traffic accidents [3]. The NHTSA estimates that 94% of serious crashes are due to human error or poor choices—including distracted driving and drunk driving [4].

The driving task necessitates interacting with the vehicle and the environment at the same time. Many body systems need to be functional to ensure the safe and timely execution of the skills required for driving [5]. Specific factors that contribute to maintaining a licence include vision, physical health and cognitive health [5]. Research indicates that cognitive abilities are important enabling factors for safe driving [6]. Research also indicates that adaptive strategies are essential to maintaining the normal parameters of driving safety in the face of illness and disability [7].

Age-related declines in the abilities of older adults provide certain obstacles to safe driving. A 2001 survey by the OECD found that 15% of those 65 or older had stopped driving, while an overwhelming number of those who continued to drive were very selective about when they did so [8]. In general, driving cessation has been linked to increasing age, socioeconomic factors, and declining function and health [9]. Negative effects of driving cessation on older adults' physical, mental, cognitive, and social functioning have been extensively studied [10–12].

Many automotive companies are developing and/or testing driverless cars. Largely, the proposed solutions follow established automation models such as the six levels of automation as defined by NHTSA [13]. Driver assistance technology presents a potential solution to problems pertaining to driver persistence and the management of fitness to drive issues in older adults. As this technology is not fully implemented and in use by the public, it is very difficult to both predict and assess its potential ethical implications and impact. Should the purpose of these systems go beyond safety? Is full automation an appropriate solution to effectively managing the apparent conflict between two goals—(1) promoting driver persistence and (2) ensuring road safety? That is, is it appropriate to enable an older driver to continue driving, even if there is a risk of a serious accident given their medical background? With crashes also comes the question of liability. Currently, lawmakers are considering who is liable when an autonomous car is involved in an accident. Such discussions raise many complex legal and ethical questions. Largely, the literature around ethics and driverless cars appears to focus on issues pertaining to (1) addressing conflict dilemmas on the road (machine ethics), (2) privacy and (3) minimising technology misuse/cybersecurity risks. These are indeed important ethical issues. However, the literature and public debate tends to avoid other serious ethical issues—specifically, issues concerning (1) the intended use and purpose of this technology, (5) the role of the person/driver (including older adult drivers) and (6) issues pertaining to the potential negative consequences of this technology.

In relation to (6), this concerns the social consequences of this technology and the potential impact on older adult identity and well-being. The future is indeed unknown. The advancement of new driving solutions raises overarching questions in relation to the values of society and how we design technology to: (a) promote positive values around ageing and enhancing ageing experience, (b) protect human rights, (c) ensure human benefit and (d) prioritise well-being. Specifically, it raises fundamental questions in relation to the value we place on promoting autonomy and social participation for older adults and optimising quality of life/well-being.

The public opinion on self-driving cars (including solutions for older adults) will determine the extent to which people will purchase and accept such systems [14]. We should not proceed with this technology just because it is available. Critically, designers must carefully consider the human dimensions of this technology and its social implications. To this end, this chapter reviews the relevant ethical considerations in relation to assisted driving solutions. Further, it presents a new ethically aligned system concept for driver assistance. In so doing, it addresses the philosophical principles that underlie the proposed driving system concept, and specifically, the role of the person.

2. Ethics, rights, digital ethics and ontological design

Ethics concerns the moral principles that govern a person's behaviour or how an activity is conducted [15]. A key distinction in ethics is the distinction between that which is unethical and that which is undesirable.

Primarily, moral principles apply to a person. However, moral code can also be ascribed to the behaviour of automated or intelligent systems (A/IS). Accordingly, driverless cars are termed 'artificial moral agents'.

The Universal Declaration of human rights (1948) enshrines all persons with human rights [16]. This includes rights pertaining to dignity (Article 1), autonomy (Article 3), privacy (Article 12), and safety (Article 29) [16]. Some would argue that rights also apply to technology and artificial agents. These are referred to as 'transhuman rights' [17, 18]. To this end, the field of roboethics has emerged. Specifically, roboethics is concerned with the moral behaviour of humans as they design, construct, use and treat artificially intelligent beings.

More broadly, 'digital ethics' or 'information ethics' deals with the impact of digital Information and Communication Technologies (ICT) on our societies and the environment at large [19]. As defined by Capurro [19], it addresses the ethical implications of things which may not yet exist, or things which may have impacts we cannot predict.

Progress is typically defined in relation to concepts of advancement and improvement. As stated by the Organization for Economic Co-Operation and Development's (OECD) 'Being able to measure people's quality of life is fundamental when assessing the progress of societies' [20]. Future technology is shaping (and will shape) our political, social and moral existence. The application of ethics to questions concerning technology development is not new. In his seminal work 'The Question Concerning Technology', the philosopher Heidegger suggests that in asking what technology is, we ask questions about who we are [21]. In so doing, we examine the nature of existence and human autonomy [21]. Such ideas have led to the concept of 'ontological design' which focuses on the 'the relation between human beings and lifeworlds' [22]. As argued by Winograd and Flores, new technology does not simply change the task, it changes what it means to be human [22]. Put simply, we are designed by our designing and by that which we have designed [23].

The Information Technology (IT) sector is taking some leaps in relation to addressing these questions. Currently, there is a large focus on issues pertaining to well-being, data privacy and cybersecurity. In 2016, Amazon, Google, Facebook, IBM, and Microsoft have established a non-profit partnership (i.e. the Partnership on Artificial Intelligence to Benefit People and Society) to formulate best practices on artificial intelligence technologies [24]. Further, the IEEE Standards Association has recently articulated a desire to create technology that improves the human condition and prioritises well-being. Specifically, the 'IEEE Global Initiative on Ethics of Autonomous and Intelligent Systems' have defined a set of core ethical principles for autonomous and intelligent systems (A/IS). As stated in 'Ethically Aligned Design (EAD1e), A Vision for Prioritizing Human Well-being with Autonomous and Intelligent Systems' [25] 'for extended intelligence and automation to provably advance a specific benefit for humanity, there needs to be clear indicators of that benefit'. Further, the IEEE Global Initiative argue that 'the world's top metric of value (Gross Domestic Product) must move beyond GDP, to holistically measure how intelligent and autonomous systems can hinder or improve human well-being' [25].

3. Well-being, identity, quality of life and self-efficacy

The concept of identity has three pillars: the person, the role and the group [26]. Personal identity refers to the concept of the self which develops over time and the life-span. This includes the aggregate of characteristics by which a person is recognised by himself/herself and others, what matters to the person and their values [27]. Crucially, autonomy is central to personal identity [27].

According to the 'Six-factor Model of Psychological Well-being', six factors contribute to an individual's psychological well-being, contentment, and happiness [28]. This includes positive relationships with others, personal mastery, autonomy, a feeling of purpose and meaning in life, and personal growth and development [28].

Quality of life is inextricably connected to well-being. As defined by the OECD, well-being can be defined/measured in relation to (1) quality of life (i.e. health status, personal security, social connection and participation/activity, work/life balance, subjective well-being, environmental quality, etc.), and (2) material conditions (i.e. income and wealth, job and earnings and housing) [29].

Self-efficacy is defined as a person's belief in his or her own ability to accomplishing a task or succeeding in specific situations. One's sense of self-efficacy can play a major role in how one approaches goals, tasks, and challenges. The promotion of self-efficacy is a key element for success in interventions designed to reduce depressive symptoms in late life [30].

4. Successful ageing

The beginning of old age is between the age of 60 or 65 [31]. Definitions of old age are multi-dimensional and include a combination of chronological, functional and social definitions [31]. Older adults are a highly heterogeneous group. Often, older adults are segmented based on factors such as ageing phases, levels of fitness, severity of physical limitations, mobility patterns and social activities. According to Rowe and Kahn, successful ageing is multidimensional, encompassing the avoidance of disease and disability, the maintenance of high physical and cognitive function, and sustained engagement in social and productive activities [32].

The prevalence of mental health issues is high in older adults as compared with the general population [30]. Older adults are at risk for developing anxiety and depression, given increased frailty, medical illnesses and medication and the potential for loss, reduced social connection and trauma (arising from injuries/accidents such as falls). On the other hand, younger older people are generally happier with a strong happiness increase around the age of 60 followed by a major decline after 75 [33].

Growth in ageing populations is a global trend. In Japan, Taiwan and Singapore, governments are defining smart ageing strategies to ensure that the growing ageing population ages well. This includes the promotion of multi-generational living, awareness of Dementia and other age-related health conditions and smart devices to monitor vital signs [34].

5. Driving task, older adult drivers and health conditions impacting on driving

5.1 Driving task

The driving is not a task isolated from everyday life. It occurs for a purpose (to get to somewhere, to see the scenery, etc.) and is often undertaken in parallel with other activities (for example, talking, listening to the radio, singing, planning-ahead and eating).

The driving task involves a complex and rapidly repeating cycle that requires a level of skill and the ability to interact with both the vehicle and the external environment at the same time [5]. Information about the road environment is obtained via the visual and auditory senses. The information is operated on by many cognitive and behavioural processes including short and long-term memory and judgement, which leads to decisions being made about driving [5]. Decisions are put into effect via the musculoskeletal system, which acts on the steering, gears and brakes to alter the vehicle in relation to the road [5]. As reported by Fuller, the overall process is coordinated via a complex process involving behaviour, strategic and tactical abilities and personality [35]. As stated in Fuller's task capability model (2005), loss of control arises when the demand of the driving task exceeds the driver's capability [35].

5.2 Older adult drivers

It is estimated that by 2030, a quarter of all drivers will be older than 65 [36]. Further, by 2030, more than 90% of men over 70 will be driving [37]. Research indicates a general increase in both car access and licensing rates in the older population [38]. This increase is mainly attributable to significant increases in the number of older female drivers [38].

A number of studies have sought to categorise older adults in terms of their physical abilities [39] their economic, geographic/spatial and activity patterns [40], use of cars as a transportation mode [41], and lifestyles and associated requirements in relation to transport services [42]. The most nuanced categorisation is that of the GOAL project which proposes five distinctive profiles or segments of older people [43]. The segments take demographics, physical and mental health characteristics, social life, living environment, mobility-related aspects and transition points into account. The five profiles differ significantly according to age and level of activity/ mobility and health [43]. They include.

- A younger and more active profile ("Fit as a Fiddle")
- A young, fit and active elderly ("Happily Connected")
- A young, severely impaired and immobile elderly ("Hole in the Heart").
- A very old, highly impaired and immobile segment ("Care-Full")
- A quite mobile and still independent senior despite his/her old age ("Oldie but a Goldie")

5.3 Older adult driving challenges

As we age, we face decisions as to whether we should (1) continue, (2) limit, or (3) stop driving. Age related declines in the abilities of older adults can be treated as obstacles/barriers to safe driving performance. These age-related changes yield specific challenges for older adults. As reported by Langford and Koppel [44], this includes:

- Psychomotor functions: joint flexibility, muscle strength, manual dexterity and coordination.
- Sensory abilities: visual acuity, contrast sensitivity, sensitivity to light, dark adaptation, visual field, space perception, motion perception, hearing.
- Cognitive abilities: fluid intelligence, speed of processing, working memory, problem solving, spatial cognition and executive functions like inhibition, flexibility and selective and divided attention.

A recent study has identified the prevalent driving errors of older adults [45]. Following a systematic review of the literature, the authors categorised the prevalent driving errors into eight categories: (1) decision-making, (2) direction and lane control, (3) lack of regulation compliance and awareness, (4) speed performance, (5) visual checking and physical control, (6) recognising and responding to signs, (7) recognising and responding to traffic lights and (8) skills involved in turning and parking. It was found that (2) direction and lane control, (1) decision-making, (7) recognising and responding to signs, and (5) visual checking and physical control were most frequent as prevalent issues for older drivers [45].

Certain unsafe driving behaviours increased in frequency as age, with drivers of 40 years or over—older people more likely to engage in driving behaviours such as (1) little or no sign of attempts to avoid dangerous driving situations, (2) lack of attention to other people and cars, (3) improper manoeuvring around curves and (4) improper or no turn signals [46].

5.4 Driver self-regulation

Self-regulation and/or compensatory behaviour of older adults is defined in relation to the tendency of older adults to minimise driving under conditions that are threatening and/or cause discomfort and conversely, to restrict their driving to conditions perceived as safe and/or comfortable [44].

Compensatory behaviour of older adults includes avoiding driving in the following situations/conditions:

- In the dark
- In bad weather
- In heavy traffic
- In new areas
- On motorways and complex road layouts
- Avoid long journeys (fatigue/tiredness)

As stated in the Eldersafe Report (2016), older road users need to be aware, acknowledge and have insight into their functional impairments in order to self-regulate [47].

5.5 Driving cessation

Health deterioration is the primary trigger/key determinant for driving cessation among older adults [48]. Medical conditions either (1) impact the fitness to drive of older drivers and/or (2) an older person's perceived fitness to drive (i.e. attitude, confidence levels, etc.). Several medical conditions and associated impairments are more prevalent in the older adult population and are, therefore, associated with ageing. These medical conditions can potentially impact the crash risk of older road users [49]. Specifically, a systematic review of the literature by Marshall identified specific conditions including: alcohol abuse and dependence, cardiovascular disease, cerebrovascular disease/TBI, depression, dementia, diabetes mellitus, epilepsy, use of certain medications, musculoskeletal disorders, schizophrenia, obstructive sleep apnoea, and vision disorders [50].

6. Self-driving cars and ethical issues

The path to automated/driverless cars began before 2000 with the introduction of cruise control and antilock brakes. Since 2000, new safety features such as electronic stability control, blind spot detection and collision and lane shift warnings have become available in vehicles. Further, since 2016, automation has moved towards partial autonomy, with features that enable drivers to stay in lane, along with adaptive cruise control technology, and the ability to self-park.

Automated driving systems are defined as systems that control longitudinal and lateral motions of the vehicle at the same time [51]. Self-driving cars use a combination of sensors, cameras, radar and artificial intelligence (AI) to travel between destinations without a human operator. The Society of Automotive Engineers (SAE) has defined six levels of driving assistance technology (level 0–5) [52].

- No automation
- Driver assistance
- Partial automation
- Conditional automation
- High automation
- Full automation

In addition, BASt [53] and the National Highway Traffic Safety Administration (NHTSA) [13] have defined equivalent standards.

Many automotive companies are developing and/or testing driverless cars. This includes Audi, BMW, Ford, General Motors, Tesla, Volkswagen and Volvo. Solutions are also being advanced by Google and Uber. As of 2019, a number of car manufacturers have reached Level 3 [54]. This level involves an automated driving system (ADS) which can perform all driving tasks under certain circumstances, such as parking the car. In these circumstances, the human driver must be ready to re-take control and is still required to be the main driver of the vehicle [54]. According to the Vienna Convention on Road Traffic (2017), as of 2017, automated driving technologies will be explicitly allowed in traffic, provided that these technologies are in conformity with the United Nations vehicle regulations or can be overridden or switched off by the driver [55].

As noted earlier, technology innovation influences societal values and raises ethical questions. As posed by BMVI, how much dependence on technologically complex systems will the public accept to achieve, in return for increased safety, mobility and convenience [56]? In relation to the advancement of assisted driving solutions, Gasser distinguishes four clusters of issues, (1) legal issues, (2) functional safety issues, (3) societal issues (including issues of user acceptability) and (4) human machine interaction (HMI) issues [53]. A recent literature review on the ethical, legal and social implications of the development, implementation, and maturation of connected and autonomous vehicles (CATV) in the United States groups the issues into the following themes: privacy, security, licensing, insurance and liability, infrastructure and mixed automation environment, economic impact, workforce disruption, system failure/takeover, safety algorithm and programming ethics, and environmental impact [57].

Largely, the literature around ethics and driverless cars appears to focus on a subset of important ethical issues. This includes issues pertaining to (1) addressing conflict dilemmas on the road, (2) privacy and protecting personal sphere, (3) minimising technology misuse and (4) the digital self and transhuman rights. In relation to (1) operational decisions have moral consequences. The issue of managing conflict dilemmas on the road poses significant challenges for autonomous vehicles. As outlined in the literature, operational decision making raises many serious questions in terms of how human life is valued. Equally, such solutions raise significant ethical questions in terms of data privacy and the sharing of sensitive/private information about a person's health condition and potential driving risk. The possibility of technology hacking is also a potential threat to the implementation of this technology. Further, issues around defining rights in the context of the augmented self (i.e. the mix of human rights and rights as apply to our digital self which is enabled/ transformed by the reach of artificial technology) are real. As argued by some, we may have to devise a set of ethics that applies to the whole continuum of our digital self and identity. Potentially, the specification of a Universal Declaration of Transhuman Rights should underpin the development of these technologies. Data gathered in a recent cross-national acceptability surveys concerning driverless vehicles indicates that the above issues are also a significant public concern [58, 59].

These are of course important both ethical and societal issues. However, the literature and public debate tends to avoid other significant issues. This includes issues pertaining to (4) the purpose and intended use of this technology, (5) issues around the role of the person/driver (including older adult drivers) and (6) the potential negative consequences of this technology, including the social consequences of this technology and its impact on well-being.

7. Research design/methodology

7.1 Objective

The high-level objective of this research was to specify the requirements for a new driving assistance system which prolongs safe driving for older adults with different ability levels, and in so doing, helps maintain cognitive and physical abilities. Importantly, the proposed system must carefully reconcile the potential conflict between (1) ensuring road safety and (2) promoting driver persistence (i.e. enabling an older driver to continue driving, even if there is a risk of a serious accident given the Drivers' medical background). From a design perspective, the challenge was to high-tech solution for users who are often averse to technology.

7.2 High level methodology

Overall, this research has involved the application of human factors methodologies to the analysis and specification of a proposed driving assistance system. Several phases of research have been undertaken. These are detailed in Appendix A. To date, this research has mostly been theoretical. Overall, the proposed driving system concept follows a multidisciplinary analysis of relevant literature pertaining to

- Older adults and positive ageing
- Segmentation of older adult drivers
- Driving task and theories of driver cessation and explanations of self-regulation
- Automated driving solutions and ethical issues
- The detection/interpretation of driver states (i.e. physical, cognitive and emotional states) using a combination of sensor-based technology and machine learning techniques
- Innovative human machine interaction (HMI) communication methods

Further, it follows the application of Human Machine Interaction (HMI) design methods including personae-based design [60] scenario-based design [61] and participatory co-design [62], to the modelling of a proposed solution. Currently, a new assisted driving solution has been defined. A preliminary workflow and multimodal communications concept has been specified in relation to several demonstration scenarios. The proposed multimodal solution will be further validated using a combination of co-design techniques and simulator evaluation.

7.3 Advancement of personae and scenarios to specify the system concept and HMI design solution

In line with a human factors approach, the proposed concept was modelled using both personae based and scenario-based design methods. Driver profiles were segmented from the perspective of driver persistence, driver health situation and ability. Overall nine driver profiles were identified. This includes:

1. Older adults in optimal health and driving as normal
2. Older adults who regulate their driving in relation to managing specific driving challenges and/or stressful (difficult) driving situations
3. Older adults who are currently driving but have a medical condition that impacts on their ability to drive
4. Continuing drivers—older adults who have continued to drive with a progressing condition—but have concerns in relation to medical fitness to drive and are at risk of giving up
5. Older adults who are currently driving and at risk of sudden disabling/medical event
6. Older adults who have stopped driving on a temporary basis
7. Older adults who have stopped driving (ex-drivers) before it is necessary
8. Older adults who have stopped when it is necessary

9. Older adults who have never driven a car (never drivers)

10. These nine profiles reflect 'ideal categories' based on the explicit project goals (safety, driver persistence, driver experience/enjoyment and health several monitoring).

These profiles were then decomposed into a series of personae. Each persona included information about the older adult's goals, their ability and health, medications, typical driving routines, typical driving behaviours and driver pain-points. For more information, please see Appendix B.

In parallel, several scenarios were defined. These scenarios followed from (1) the project goals (i.e. top down approach) and, (2) specific driving challenges and older adult driver behaviours, as identified in the literature review (i.e. bottom up approach). These include:

1. Driver is enjoying drive—everything going well

2. Driver is distracted by their mobile phone ringing

3. Driver feels stressed given traffic delays

4. Driver has taken pain medications and is drowsy

5. Driver is fatigued after long day minding grandchildren

6. Driver is having difficulty parking (visual judgement)

7. Sudden advent of acute medical event

8. Driver is having difficulty remembering the correct route

9. Driver has taken alcohol and is over the legal limit

As indicated in **Table 1**, the different scenarios were classified in terms of interpretation challenges.

Following this, the scenarios were associated with specific user profiles and personae (see **Table 2**).

Lastly, the specific scenarios were further decomposed in relation to (1) a time sequence/text narrative, (2) the sensing framework and behaviour of sensor technology and machine learning, and (3) multi-modal communications.

8. Key findings/results

8.1 Segmenting older adult drivers and role of new technology

Nine end user profiles have been identified—see **Table 3**. Specific system goals/requirements are associated with different profiles. It is suggested that the proposed solution might target profiles 1–7, and potentially profile 9.

8.2 Driving scenarios and ethical issues

The different driver scenarios as defined in **Table 1** raise a myriad of ethical questions—in addition to legal issues and issues pertaining to societal/user acceptability. For example,

Interpretation challenge		Explanation of the interpretation challenge	Scenario examples
1	Task support/feedback	Addresses driving challenges and typical supports required	Parking support Navigational assistance Assistance changing lanes
2	Activation/"flow"	Incorporates multiple psychological states: stress/anger/excitement/workload/engagement including driver difficulties and driver behaviour	Flow/enjoying drive Stress given traffic delays Intelligent driving
3	Distraction and concurrent task management	Addresses age-related cognitive and perceptual challenges including driver difficulties and driver behaviour	Distraction from mobile phone ringing Talking with passenger/checking GPS directions and driving
4	Fatigue and drowsiness	Many medical conditions and drugs also manifest this way	Fatigue
5	Intoxication—alcohol/drugs/related medical conditions	Other drugs and some medical conditions manifest similarly	Alcohol Prescription drugs
6	Heart attack/stroke	Addresses fear factor—which may discourage older drivers from driving	Heart attack Stroke

Table 1.
Interpretation challenges and scenarios.

Interpretation challenge		Scenario	Profile	Personae
1	Task support/feedback	Driver needs assistance with parking	2. Older adults who regulate their driving in relation to managing specific driving challenges and/or stressful (difficult) driving situations (perceived safety risk or complexity)	Mary
2	Activation/flow	Flow	4. Continuing drivers: older adults who have continued to drive with a progressing condition, but have concerns in relation to medical fitness to drive and are at risk of giving up	Sarah/James
		Stress	5. Older adults who are currently driving and at risk of sudden disabling/medical event	Louise
		Intelligent driving	2. Older adults who regulate their driving in relation to managing specific driving challenges and/or stressful (difficult) driving situations (perceived safety risk or complexity).	Mary
3	Fatigue and drowsiness	Fatigue	1. Older adults in optimal health and driving as normal	Elizabeth/Sam
4	Distraction and concurrent task management	Distraction	2. Older adults who regulate their driving in relation to managing specific driving challenges and/or stressful (difficult) driving situations (perceived safety risk or complexity)	Tom
		Concurrent Task Management	3. Older adults who are currently driving but have a medical condition that impacts on their ability to drive	Richard
5	Intoxication	Alcohol	1. Older adults in optimal health and driving as normal	James
		Prescription drugs	5. Older adults who are currently driving and at risk of sudden disabling/medical event	Rory
6	Heart attack/stroke	Heart attack	5. Older adults who are currently driving and at risk of sudden disabling/medical event	Brian
		Stroke	5. Older adults who are currently driving and at risk of sudden disabling/medical event	Louise

Table 2.
Interpretation challenges, scenarios, user profiles and personae.

- How is the human role and well-being being considered in relation to the development of these systems?

- What is the role of older adult and what level of choice do they have in relation to mode of operation?

- What level of impairment is acceptable for an older driver to keep driving?

#	User profile	Goals/role of new technology
1	Older adults in optimal health and driving as normal.	Driving enabling life-long mobility Monitor driver's task and driver's capability Monitor driver states that impact on driver capability and provide task assistance to ensure safety Promote confidence for older driver Promote comfortable, enjoyable and safe driver experience
2	Older adults who regulate their driving in relation to addressing specific driving challenges	As (1) and... Technology directly addresses causes of self-regulation
3	Older adults who are currently driving but have a medical condition that impacts on their ability to drive	As (1) and... New car directly addresses challenges associated with condition Monitor driver state in relation to specific medical condition, and provide task assistance to ensure safety
4	Continuing drivers—older adults who have continued to drive with a progressing condition—but have concerns in relation to medical fitness to drive and are at risk of giving up	As (1) and... New tech might monitor conditions and provide feedback—continue with licence/evidence, keep safe
5	Older adults who are currently driving and at risk of sudden disabling/ medical event	As (1) and... New tech might monitor conditions and provide feedback New tech might take relevant action based on detection of onset of medical event
6	Older adults who have stopped driving on a temporary basis	As (1) and... Monitor driver state and health condition and provide task assistance to optimise safety
7	Older adults who have stopped driving (ex-drivers) before it is necessary	As (1), (2), (3), (4) and (5)
8	Older adults who have stopped when it is necessary	N/A
9	Older adults who have never driven a car (never drivers)	As (1) and... Motivate to buy car/learn to drive, given protections provided by new car and associated driver experience

Table 3.
User profiles and goals.

- Should the system determine the level of automation/assistance, or the older adult?
- Should the driver be able to take control of the car at any point?
- How is information about the health status of the driver, their driving challenges, driving routines and any driving events being stored?
- Who has access to driver profiles, health information and incident information?

For a full list of issues, please see Appendix C.

Overall, there is much overlap between ethical issues and legal issues. There is also much commonality between ethical issues and user acceptability/societal issues. Further, many of the ethical and societal/user acceptability issues are also HMI/human factors issues (for example, handover of control and role of the older adult in the system, etc.).

In principle, ethical issues and issues concerning societal/user acceptability pertain to all profiles as defined previously. Critically, these ethical issues have meaning in the context of different degrees of automation. Some issues pertain to the specific level of driving automation (i.e. manual, partially automated/function specific, highly automated, fully automated), while others present to all.

8.3 Framing the design problem and system objectives

The design problem is framed in relation to advancing systems that can detect the health and psychological/emotional condition of the driver, so that the vehicle responds as appropriate, while also ensuring a positive/enjoyable driving experience and promoting driver self-efficacy.

To this end, three high level goals for the system have been defined. These are:

1. Safe driving for older adults

2. Driver persistence

3. Positive driver experience

Accordingly, the requirement is to advance a system which can detect the health and psychological/emotional condition of the driver so that the vehicle responds as appropriate (i.e. promoting engagement/alertness, providing task supports, taking over the driving task if the driver is impaired and/or calling an ambulance).

8.4 Refining system goals: human benefit and well-being (objectives and measures)

It is very difficult to both predict and assess the potential ethical implications and impact of this technology. However, we can document key performance indicators (KPIs) relevant to the potential success of this technology once it is introduced and used by the public.

As stated previously, we have defined three high level goals for the system. These goals have been reformulated in terms of objectives concerning human benefit and well-being and associated measures/KPI's. These are described in **Table 4**. As indicated in **Table 4**, there is a relationship across goals (1), (2) and (3), and the associated objectives and metrics.

#	System goal	Human benefit and well-being objectives/ targets (design outcomes)	Metric (outcome indicators)
1	Safe driving for older adults	Driver feels safe Driver feels in control The car is in a safe state	Subjective perception of safety/security Objective measure of car safety (position on road/lane, speed)
2	Driver persistence	Car as an enabler of active ageing/positive ageing—and allied health benefits Car contributing to eudaemonia (living well) Car contributing to a sense of having a purpose Car as an enabler of mobility Supporting social connection and participation Supporting citizenship, etc.	Health status Mobility status Positive human functioning and flourishing Social capitol Personal growth
3	Driver experience	Driver feeling happy/enjoying driving activity Emotional state/psychological well-being (avoidance of stress) Driver in control Focus on ability (available capacity) Promote adaptation and bricolage	Subjective enjoyment of driving Subjective feeling of human agency/ independence Subjective well-being

Table 4.
System goals, well-being objectives and well-being metrics.

9. Proposed co-pilot/adaptive automation driving solution

9.1 High level principles underlying system concept

The third phase of research involved the specification of the high-level system logic and associated principles associated with this concept. The high-level principles associated with the system logic are grouped into six themes as follows:

1. Philosophy of the system

2. Technology and the conceptualization of the driver

3. Technology and the conceptualization of the driver task and driving experience

4. Driver health conditions and emotional/psychological State

5. Detecting symptoms with sensors

6. Using multi-modal technology to promote safe driving and a positive driving experience

As indicated in **Figure 1**, the principles associated with (1) are derived from related principles relating to (2), (3), (4), (5) and (6). In addition, the principles related to (5) follow from an understanding of (4) and feed into (2) and (3) and so forth. Subsequent sections focus on principles related to (1) and (2).

9.2 Philosophy of the system

9.2.1 Assistance/adaptive automation (balancing safety and driver persistence/quality of life)

The proposed co-pilot system carefully reconciles the potential conflict between two goals—(1) ensuring road safety and (2) promoting driver persistence (i.e. enabling an older driver to continue driving, even if there is a risk of a serious accident given the drivers' medical background). Overall, the technology is designed to provide different levels of assistance/automation to drivers so that accidents are avoided (i.e. safety). Three levels of assistance are proposed.

1. No response—all seems to be in order, the driver is alert and attentive, driving well; there is no basis for an intervention

2. Driving assistance—one or more driver factors have been identified; they are not an immediate threat, but the driver could do with some assistance to drive safely and/or manage their own emotions. Driving assistance could take a range of forms:

    - An alert to the driver

    - Adjusting car settings

    - Auto-braking/speed reduction

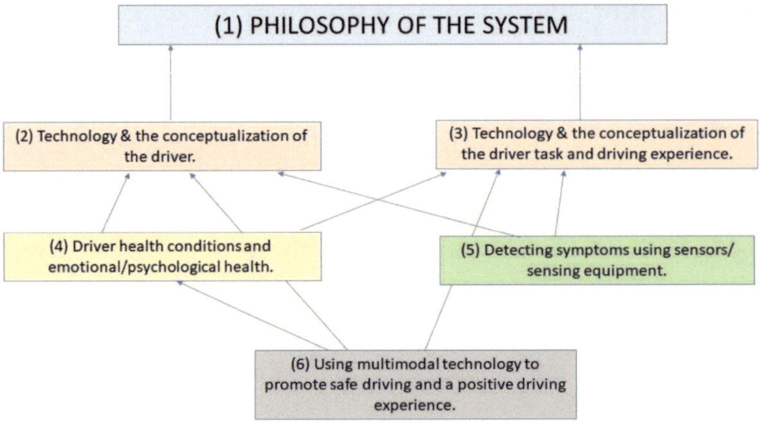

Figure 1.
High level principles.

- Temporary co-pilot in charge
- Task assistance
- Task information

3. Safety critical intervention—the driver's health and/or safety are at immediate risk; the co-pilot needs to make a strong intervention. This could include:

- Auto-park and engine stop
- External warnings to other road users
- Alerts to emergency services

To this end, we are proposing assistance (i.e. adaptive automation) and not full automation. Normally, the older adult driver chooses the level of task assistance required. However, the system also recommends different levels of assistance based on the driver's profile (level of ability), and real time context (i.e. driver state and driver behaviour). In particular circumstance, if the system detects that (1) the driver is in a seriously impaired state (i.e. alcohol or medications), (2) there is a potential for a safety critical event, or (3) the driver is incapacitated, then authority moves to 'automation'. Accordingly, the proposed co-pilot system is both reactive and predictive.

9.2.2 Universal design

The system is designed to be usable, accessible, and understood by people of all ages with different abilities and health conditions. To this end, the system/co-pilot system provides three levels of assistance, taking into account the diverse driving situations and needs of different drivers (including older adult drivers).

9.2.3 Positive ageing and self-efficacy

The proposed co-pilot system is premised on concepts of successful/positive ageing and self-efficacy. Although certain conditions occur in old age (and impact on the driving task), old age itself is not a disease. Ageing (and the associated changes in functional, sensory and cognitive function) is a normal part of life. To this end, the system seeks to normalise ageing, and not treat ageing as a 'problem' or 'disease'. The driving solution (i.e. car, sensor system, co-pilot and HMI) is designed to optimise the abilities and participation of older adults. That is, it addresses what older adults can do as opposed to focusing on declining capacities.

9.2.4 Ability, adaption and assistance (not automation)

The co-pilot is conceptualised as a means/intervention to ensure that older adults drive safely and for longer. Critically, the technology supports continued and safe driving for all adults, including those adults at risk of limiting their driving and/or giving up. Accordingly, concepts of ability, adaption and assistance (as opposed to vehicle automation) underpin the system logic. To achieve this, the proposed technology provides different levels of assistance, tailored to the older adults (1) ability, (2) health and (3) the real-time physical and psychological/emotional health. In general, this will deliver benefits for the wider population and not just older adults.

9.2.5 Interpretation of driver profile and real-time context

The ability of the driver to perform the driving task depends on the driver's ability (i.e. functional, sensory and cognitive), his or her driving experience and the 'real time' state of the driver (i.e. health, level of fatigue, emotional state, etc.) and the operational context (i.e. cabin context, road context, weather and traffic). Thus, to provide targeted task support to the driver, the system combines (1) an understanding of the driver's profile (i.e. ability and driving experience) and (2) an interpretation of the real time context (i.e. the state of the driver and the operational context).

9.2.6 Focus on interpretation challenges and not conditions/state

The critical objective for the system is not to precisely diagnose the drivers' condition/state but to interpret the implications for the driving task and the driver. According, the driving assistance system logic addresses 'interpretation challenges' rather than the driver condition or state. This is achieved in relation to six high-level interpretation challenges. These include.

1. Task support/feedback

2. Activation/flow

3. Distraction and concurrent task management

4. Fatigue and drowsiness

5. Intoxication

6. Heart attack/stroke

9.2.7 A learning system will enable driver persistence and a positive driver experience

Underpinning the system logic, is a vision of the co-pilot as a learning system. Arguably, a human-centric design philosophy necessitates continuous learning on the behalf of the co-pilot (i.e. including AI/machine learning). If the co-pilot can learn about those situations and tasks that prove challenging and/or stressful for the older adult driver (i.e. driving in traffic, poor visibility, changing lanes, parking and so forth, etc.), then it can truly tailor the task support that it provides to the driver. This tailored task support is predictive/intelligent, ensuring that the driver persists in challenging driving situations, while also enjoying their drive.

9.3 Technology and the conceptualization of the driver

9.3.1 Role of driver in the system and adaptive automation

The proposed system maintains the autonomy of the individual. In principle, the driver is able to choose (and/or switch off) task support and advanced levels of automation, if they so choose. Overall, we are starting from the point of the engaged driver, who has capacity and ability. In this way, the system supports a vision of the older adult driver as 'in control'. The role of the driver is to work in partnership with the 'co-pilot', to achieve a safe and enjoyable drive. Critically, the system treats the driver as 'capable' and 'in charge' unless it detects that the driver is incapacitated and/or there is a potential for a safety critical event (i.e. level 3 assistance/safety critical intervention). If the system detects that the driver is in a seriously impaired state and/or incapacitated, or that a safety critical event is imminent, then the principle of 'driver autonomy' is outweighed by that of safety. In such cases, authority moves to 'automation'.

9.3.2 Driver as a person (holistic approach)

The proposed driving assistance system is premised on a conceptualisation of the driver/older adult as a person and not a set of symptoms/conditions (i.e. holistic approach). Specifically, biopsychosocial concepts of health and wellness inform the logic of the proposed driving assistance system. The system is concerned with all aspects of the driver's wellness, including the driver's physical, social, cognitive and emotional health.

9.3.3 Diversity in older adult population

Critically, the driving assistance system logic is premised on the idea that all older adult drivers are not the same. Older adult drivers vary in many ways including body size and shape, strength, mobility, sensory acuity, cognition, emotions, driving experience, driving ability (and challenges) and confidence. In relation to driving situation and ability, we have segmented older adults into the following high-profiles or clusters—as indicated previously. These profiles have been further specified in relation to a series of personae. Critically, the system logic directly addresses the needs and requirements of these specific personae.

9.3.4 Upholding rights (autonomy, dignity and privacy)

The acceptability of the proposed system largely depends upon how it treats certain issues pertaining to driver rights. Overall this technology is designed to uphold an older adult's rights. This is specifically salient in relation to preserving driver autonomy, monitoring the driver state and recording driver health information. As outlined earlier, the technology maintains the autonomy of older adults (i.e. the starting point is the engaged driver). Further, we are proposing that information captured about the person's current health and wellness and driving challenges/events is NOT shared with other parties. In all cases, the driver is in charge of their own data and decisions about how it is stored and shared with others.

10. Discussion

10.1 Ontological design, digital ethics and coping with change

As highlighted by Fry, the introduction of new technology has the potential to transform what it means to be human [23]. In this way, the introduction of new assisted driving solutions presents a challenge to our being. Design decisions are normative—they reflect societal values concerning human agency and human identity/avoiding ageism. In particular, they provide an opportunity to foster quality of life for older adults as they age, and to promote positive ageing. Design/technology teams thus exercise choice in relation to what is valued and advancing technology that improves the human condition (and not worsens it).

The discovery and utilisation of fire by early humans was of course transformative and positive [63]. It shaped how we eat, kept warm and how we protected ourselves. However, less examined are the negative by-products that came with fire, and the ways in which humans may or may not have adapted to them [63]. In the same way, it is important that designers consider issues pertaining to potential technology impact in terms of the three strands of health and wellness (i.e. biological, psychological and social health). In particular, designers should consider protections concerning the 'unknown' future implications of this technology (including the potential negative social consequences).

In relation to the introduction of other consumer and information technologies (for example, mobile phones and social media), many important questions were posed 'post hoc'. As stated by Heraclitus, 'One cannot step twice in the same river' [64]. These technologies have resulted in many changes to previously established social norms. Arguably, social norms in relation to identity and privacy and associated information sharing, have appeared to change—and without serious questioning of the implications of this. Further, in its early stage, designers need not properly consider the potential social consequences of this technology (for example, social isolation and depression).

Nonetheless, just because the horse has bolted (i.e. the automotive industry is currently advancing and testing driverless cars), does not mean there is nothing to be achieved and/or that we are powerless. As mentioned previously, the availability of this technology does not mean that we have no choice. Critically, we need to challenge existing design assumptions from the perspective of human benefit, well-being and rights. In this regard, the IEEE Global Initiative represents a positive step in this direction.

Salganik proposes a hope-based and principle-based approach to machine ethics [65]. This is contrasted with a 'fear-based and rule-based' approach in Social Science, and a more 'ad hoc ethics culture' as emerging in data and computer science [65]. Hope is not enough! As evidenced in this research, principles need to be both articulated and then embedded in design concepts. Importantly, human factors methods are useful here—in relation to considering different stakeholders and adjudicating between conflicting goals/principles.

10.2 System purpose and human benefit

In line with what is argued by the IEEE, A/IS technologies can be narrowly conceived from an ethical standpoint. Such technologies might be designed to be legal, profitable and safe in their usage. However, they may not positively contribute to human well-being [25]. Critically, new driving solutions should not have 'negative consequences on people's mental health, emotions, sense of themselves, their autonomy, their ability to achieve their goals, and other dimensions of well-being' [25].

Arguably, as demonstrated in this research, we can define an ethically aligned design in relation to several key concepts. This includes (1) human role, (2) human benefit, (3) rights, (4) progress and (5) well-being. These concepts provide structuring principles to guide the design of new driving assistance systems.

A key theme of this research has been about defining the purpose and role of new driving assistance technologies. As designers we decide what ethical guidelines AI in autonomous vehicles will follow. The analysis of relevant health literature and TILDA data has identified specific conditions that impact on older adult driving ability [66]. As such, it has provided an empirical basis for addressing ethical dilemmas around whether full automation is an appropriate solution to effectively managing the conflict between two goals—namely, (1) promoting driver persistence and (2) ensuring road safety. It is argued that the three levels of driver assistance represent an ethically aligned solution to enabling older drivers to continue driving, even if there is a risk of a serious accident given their medical background. Evidently, some medical conditions do not negatively impact on safe driving. However, there are other conditions that pose challenges to safe driving, and others still that make it unsafe to drive. The proposed solution is designed to directly address this fact—to promote driver persistence and enablement in these different circumstances, albeit while simultaneously maintaining safety.

Human benefit is an important goal of A/IS, as is respect for human rights. In terms of rights, this includes the rights of (1) older adult drivers and (2) other road users and pedestrians who may be negatively affected by older adult driving challenges and specifically, health events such as strokes and heart attacks. The specification of benefits is not straightforward. People benefit differently. Also, benefits are not always equal for all people, as driving system that benefits older adults must also benefit other road users and pedestrians. In this way, the proposed system must be verifiably safe and secure. We must ensure the safety of all drivers and pedestrians. Benefits in relation to older adult mobility must not outweigh safety concerns (i.e. we cannot address benefit from a narrow perspective/prioritise one stakeholder).

10.3 Design problem and ethical vision: enablement and positive ageing

The design problem—prolonging safe driving for older adults is framed in relation to a philosophy of 'enablement' and positive models of ageing. Crucially, the proposed vision of 'technology progress' in closely intertwined with concepts of progress from a societal values perspective. The proposed co-pilot system is premised on concepts of successful/positive ageing and self-efficacy. The system seeks to normalise ageing, and not treat ageing as a 'problem' or 'disease'. The driving solution (i.e. car, sensor system, co-pilot and human machine interface) is designed to optimise the abilities and participation of older adults. That is, it recognises what older adults can do as opposed to focusing on declining capacities. Further, the co-pilot is conceptualised as a means/intervention to ensure that older adults drive safely and for longer. The proposed technology supports continued and safe driving for all adults, including those adults at risk of limiting their driving and/or giving up when there is no medical/physical reason for doing so.

Arguably, existing high automation approaches do not support positive ageing. Crucially, 'technology progress' in closely intertwined with concepts of progress from a societal values perspective. New assisted driving solutions provide an opportunity to change/improve the lived experience of older adults, particularly in relation to autonomy and social participation. Enabling driver persistence is an issue for all of society, not just older adults.

10.4 Personalisation and role of AI

Many negative driving experiences are linked to frustrations with the vehicle not being configured for the driver. Drivers are highly diverse in terms of size, strength, angle of vision and experience of different vehicles. Older drivers present even greater diversity when limitations of movement, hearing, eyesight, memory emerge. It is argued that personalisation is central to fostering a positive driver experience. For example, vehicle sensors can be used to detect which driver is driving and to adjust the vehicle parameters accordingly (i.e. angle of mirrors, steering wheel, seat, etc.). Moreover, personalisation offers an enormous opportunity to ensure that task support and multimodal feedback is configured according to knowledge of the particular driver's ability (including sensory ability), driving routines and routes and typical challenges/errors.

A human-centric and ethically aligned design philosophy necessitates continuous learning on the behalf of the assistance system (i.e. including AI/machine learning). If the assistance system can learn about those situations and tasks that prove challenging and/or stressful for the older adult driver (i.e. driving in traffic, poor visibility, changing lanes, parking and so forth, etc.), then it can tailor the task support that it provides to the driver. This tailored task support is predictive/intelligent, ensuring that the driver persists in challenging driving situations, while also enjoying their drive.

10.5 Role of human factors

New technology raises complex ethical questions. Assessing the ethical implications of things which may not yet exist, or things which may have impacts we cannot predict, is very difficult. However, this should not be barrier to posing important questions and ensuring that these questions are addressed as part of the design process. Typically, the human factors discipline is concerned with issues around intended use, user interface design and technology acceptability. As demonstrated in this research, human factors research should extend its remit to include examination of ethical issues pertaining to new technology, and specifically, how well-being, rights and human value/benefit should be considered in terms of design solutions. In this way, HF methods can be used to provide some protections to ensure that ethical issues are considered. As demonstrated in this research, the application of a personae/scenario-based design approach allows us to consider the ethical dimension of these technologies. Further, the translation of system objectives in relation to well-being and human benefit objectives and associated metrics—ensures that well-being and human benefit is both a reference point and a design outcome. We may not have certainty as regards potential future technology impact, but at least we are asking important questions so as to pave the way for an ethically aligned technology of which well-being and human value is a cornerstone. The design and implementation of ethically aligned technology takes leadership and education. It also requires adopting a multi-disciplinary perspective and ensuring diverse disciplines are involved in solution design (including persons trained in ethics and moral reasoning). Further, a crucial element of the design process to ensure an ethical product is rigorous experimentation in a simulator using a co-design approach.

10.6 Next steps

The initial concept requires further elaboration and specification. In line with a human factors approach, a series of co-design and evaluation sessions will be undertaken with end users. In addition, the proposed solution will be evaluated in using a driving simulator. A health event cannot be induced as part of a driving simulation exercise. However, we can evaluate the overall concept, driver responses and the usability of specific driver input/output communication mechanisms.

11. Conclusions

The proposed design/automation approach reflects an ethically aligned and principled approach to a multi-dimensional design problem. Human benefit, well-being and respect for human rights and identity are important goals for new assisted driving technologies. Such systems must also be verifiably safe and secure. In this way, the solution needs to carefully balance goals around safety and human benefit. As indicated in this research, well-being and human benefit goals and associated KPI are defined to ensure that these concepts are properly considered in the design process, and to ensure that well-being and human benefit is a tangible outcome of new assisted driving solutions.

Arguably, existing high automation approaches do not support positive ageing. Crucially, 'technology progress' in closely intertwined with concepts of progress from a societal values perspective. New assisted driving solutions provide an opportunity to change/improve the lived experience of older adults, particularly in relation to autonomy and social participation. Enabling driver persistence is an issue for all of society and not just older adults.

The application of new car-based sensors underpinned by machine learning techniques, and innovative multimodal HMI communication methods can support driver persistence, driver enablement and successful ageing. The proposed adaptive automation/co-pilot concept is predicated on an analysis of the literature and relevant ageing data (i.e. TILDA data). The co-pilot concept and associated innovative multimodal HMI will be further elaborated using human factors/stakeholder evaluation methods (for example, participatory co-design and evaluation in a test simulator).

It is anticipated that this new car-based technology will deliver (1) safe driving (2) driving persistence and (3) an enhanced driver experience. (4) Health monitoring is built into (1), (2) and (3). In this way, health monitoring is not a goal of new driving assistance systems. Rather, it is an enabler of driver assistance systems and promotes safe driving, driving persistence and an enhanced driver experience.

Acknowledgements

The authors would like to thank the Science Foundation Ireland (SFI) who co-sponsored this research.

Conflict of interest

The authors declare no conflict of interest.

Appendices and Nomenclature

A. Research phases and status

See **Table 5.**

Phase	Description	Details	Status
1	Literature review	Driver task, older adult driver segmentation, older driver challenges, self-regulation of driving, driver cessation	Complete
		Successful ageing	
		Health conditions that impact on older adult driving	
		Assisted driving concepts and issues pertaining to ethics and user acceptability	
		The detection/interpretation of driver states (i.e. physical, cognitive and emotional states) using a combination of sensor-based technology and machine learning techniques	
		Innovative multimodal communication approaches and driving solutions	
2	Advancement of profiles, personae and scenarios	Segmentation of driver profiles in relation to driver persistence and ability	Complete
		Advancement of personae and scenarios	
3	Specification of theoretical principles underpinning advancement of new driving concept	Advancement of technology role, purpose and approach (adaptive automation)	Complete
4	Specification of high-level multimodal HMI approach	Specification of scenarios Iterative refinement of scenarios and multimodal concept Iterative integration of scenarios with sensor and machine learning research	Complete
5	Co-design of evaluation of HMI concept	Specification of preliminary UI concept Preliminary co-design/evaluation with stakeholder panel (desktop simulation of high-level concept	Ongoing
6	Simulator evaluation	Detailed evaluation in simulator	To do

Table 5.
Research phases and status.

B. Personae

See **Figure 2** and **3.**

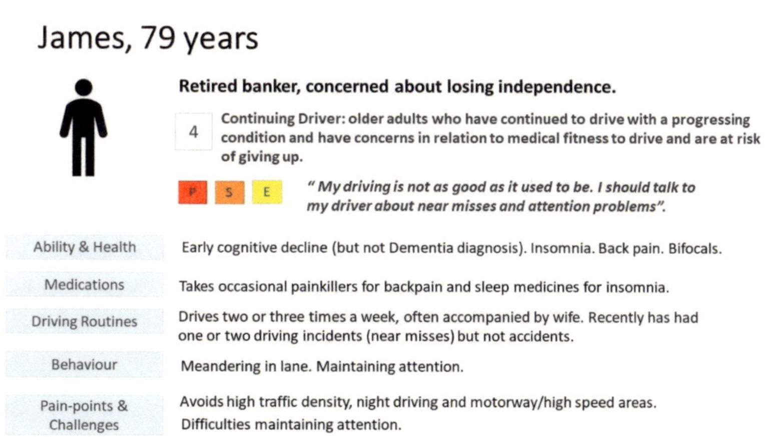

Figure 2.
Personae (James).

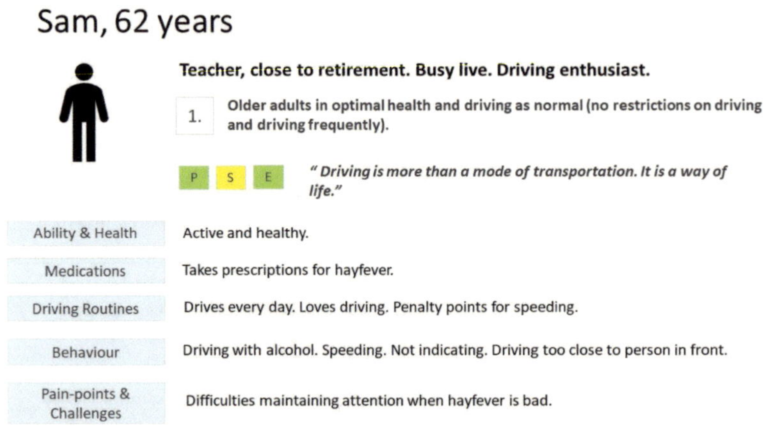

Figure 3.
Personae (Sam).

C. Summary of ethical, legal and societal/user acceptability issues

See **Table 6**.

#	Question/issue	Keywords
1	How much dependence on technologically complex systems (potentially based on artificial intelligence with machine learning capabilities) will the public accept to achieve, in return, more safety, mobility and convenience?	Ethics, user, societal acceptability
2	Agreeing/defining the purpose and role of these systems? Should the purpose go beyond safety?	Ethics, user, societal acceptability, safety
3	Agreeing/defining the role of the individual in the system	Ethics, user, societal acceptability, legal
4	Dealing with conflict between two goals—promoting driver persistence and ensuring road safety (enabling an older driver to continue driving, even if there is a risk of a serious accident given medical background)	Ethics, user, societal acceptability, legal, safety, driver persistence
5	Should the system determine the level of automation/assistance, or the older adult? Is this something that the older adult chooses (and can modify in real-time), or is it prescribed given profile information?	Ethics, user, societal acceptability, legal, safety, driver persistence
6	What is the intended use? Are these reactive and/or predictive systems?	Ethics, user, societal acceptability, legal, HF
7	Balancing expected benefits versus risk (system failure, hacking, etc.)	Ethics, user, societal acceptability, legal, HF
8	What are the legal obligations of the driver, if the driver is taken out of the loop (i.e. full automation)?	Ethics, legal, societal/user acceptability
9	Who is to blame if there in accident—the driver or the co-pilot?	Ethics, legal, societal/user acceptability
10	If the driver is in an impaired state (i.e. Alcohol, drug use, medications) should they be allowed driver only if automation take control? What level of impairment is acceptable?	Ethics, legal, societal/user acceptability
11	Addressing conflict dilemmas on the road? How should the car act (what aught the automated car do/decision logic), in cases where a choice must be made between one of two evils (decision between one human life and another)?	Ethics, legal, societal/user acceptability, safety

#	Question/issue	Keywords
12	In what circumstances, can automation take control over the car (override the decisions of the driver)?	Safety, human factors, legal, ethics, user/societal acceptability
13	Should the driver be able to take control of the car at any point? Should the driver always be in control? What tasks are suitable to delegate to automation?	Safety, human factors, legal, ethics, user/societal acceptability
14	Protection of the personal sphere? User control over own information? Information span personal profile, health profile, location tracking, destination tracking, safety behaviour, etc.	Legal, ethics, user/societal acceptability
15	Handover issues/transition of control (human to technology handover and tech to human, etc.)	Safety, human factors, ethics, user/societal acceptability
16	Software hack and misuse Cybersecurity threats and vulnerabilities—both in relation to personal information and car security	Safety, human factors, ethics, user/societal acceptability
17	Safety issues related to equipment or system failure. System/equipment failure and vehicle performance in unexpected situations	Safety, human factors, ethics, user/societal acceptability
18	Acceptable levels of workload—monitoring automation status.	Safety, human factors, user acceptability
19	Personality traits and assisted driving	Safety, societal acceptability, ethics
20	Dealing with emotions and providing feedback to the driver	Health monitoring, safety, user/societal acceptability, ethics, legal
21	Does the system provide the driver with feedback about their health?	Health monitoring, safety, user/societal acceptability, ethics, legal
22	System and consideration of information available to potential passengers?	Safety, driver experience, ethics, legal, user/societal acceptability
23	Environmental implications	Legal, user/societal acceptability
24	Training required—changes to existing driver training?	Safety, legal
25	Recording of information for crash analysis purposes? Similar to cockpit voice recorder and flight data recorder?	Safety, ethics, legal, user/societal acceptability
26	Should self-vehicles be able to operate in normal traffic or in separate lanes?	Driver experience, ethics, legal, user/societal acceptability
27	Data transmission? Sharing of information with other parties?	Ethics, legal, user/societal acceptability
28	Whether drivers expect to find it enjoyable or not? Should it be enjoyable?	Driver experience
29	Should self-driving vehicles be able to move while unoccupied?	Ethics, safety, driver experience
30	How should self-driving vehicles interact with other non-self-driving vehicles?	Ethics, safety, driver experience

Table 6.
Ethical, legal and societal/user acceptability issues.

Author details

Joan Cahill[1*], Katie Crowley[1], Sam Cromie[1], Alison Kay[1], Michael Gormley[1], Eamonn Kenny[2], Sonja Hermman[2], Ciaran Doyle[3], Ann Hever[4] and Robert Ross[5]

1 School of Psychology, Trinity College Dublin (TCD), Ireland

2 ADAPT Centre, Trinity College Dublin, Ireland

3 Department of Physical Education and Sport Sciences, University of Limerick, Ireland

4 The Irish Longitudinal Study on Ageing, TCD, Ireland

5 School of Computing, Dublin Institute of Technology, Ireland

*Address all correspondence to: cahilljo@tcd.ie

References

[1] Webber SC, Porter MM, Menec VH. Mobility in older adults: A comprehensive framework. The Gerontologist. 2010;**50**:443-450. DOI: 10.1093/geront/gnq013

[2] United Nations. World Population Prospects. 2017. Available from: https://esa.un.org/unpd/wpp/Publications/Files/WPP2017_KeyFindings.pdf

[3] World Health Organization (WHO). Global Status Report on Road Safety 2018. Available from: https://www.who.int/violence_injury_prevention/road_safety_status/2018/en/

[4] National Highway Traffic Safety Administration. National Motor Vehicle Crash Survey Results. 2008. Available from: https://crashstats.nhtsa.dot.gov/Api/Public/ViewPublication/811052

[5] Road Safety Authority. Medical Fitness to Drive Guidelines. 2016. Available from: http://www.rsa.ie/Documents/Licensed%20Drivers/Medical_Issues/Sláinte_agus_Tiomáint_Medical_Fitness_to_Drive_Guidelines.pdf

[6] Anstey KJ, Wood J, Lord S, et al. Cognitive, sensory and physical factors enabling driving safety in older adults. Clinical Psychology Review. 2005;**25**:45-65

[7] Langford J, Braitman K, Charlton J, Eberhard J, O'Neill D, Staplin L, et al. TRB Workshop 2007: Licensing authorities' options for managing older driver safety practical advice from the researchers. Traffic Injury Prevention. 2008;**9**(4):278-281

[8] Organization for Economic Co-Operation and Development. Ageing and Transport, Mobility Needs and Safety Issues. Paris, France: OECD Publications; 2001

[9] Marottoli RA, de Leon CFM, Glass TA, Williams CS, Cooney LM Jr, Berkman LF. Consequences of driving cessation: Decreased out-of-home activity levels. Journal of Gerontology. 2000;**55**(6):S334-S340

[10] Ackerman ML, Edwards JD, Ross LA, Ball K, Lunsman M. Examination of cognitive and instrumental functional performance as indicators for driving cessation risk across 3 years. The Gerontologist. 2008;**48**(6):802-810

[11] Curl A, Stowe J, Cooney T, Proulx C. Giving up the keys: How driving cessation affects engagement in later life. The Gerontologist. 2014;**54**(3):423-433

[12] Edwards JD, Lunsman M, Perkins M, Rebok GW, Roth DL. Driving cessation and health trajectories in older adults. The Journals of Gerontology. Series A, Biological Sciences and Medical Sciences. 2009;**64**(12):1290-1295. DOI: 10.1093/gerona/glp114

[13] National Highway Traffic Safety Administration. Preliminary statement of policy concerning automated vehicles. 2013. Available from: http://www.nhtsa.gov/staticfiles/rulemaking/pdf/Automated_Vehicles_Policy.pdf

[14] Kyriakidis M, Happee R, De Winter JCF. Public opinion on automated driving: Results of an international questionnaire among 5,000 respondents. Transportation Research Part F: Traffic Psychology and Behaviour. 2015;**32**:127-140

[15] Oxford English Dictionary. Definition of Ethics. 2019. Available from: https://en.oxforddictionaries.com/definition/ethics

[16] United Nations. Universal Declaration of Human Rights. 1948. Available from: https://www.un.org/en/

universal-declaration-human-rights/index.html

[17] Bostrom N. "The transhumanist FAQ: A general introduction." Version 2.1. World Transhumanist Association; 2013. Available from: http://www.transhumanism.org/resources/FAQv21.pdf

[18] Evans W. Posthuman Rights: Dimensions of Transhuman Worlds. Universidad Complutense, Madrid: Teknokultura; 2015. Available from: https://revistas.ucm.es/index.php/TEKN/article/view/49072

[19] Capurro R. Digital Ethics. The Academy of Korean Studies (ed.): 2009 Civilization and Peace, Korea: Academy of Korean Studies; 2010. pp. 203-214

[20] Organization for Economic Co-Operation and Development OECD 2. Nd World Forum—Istanbul 2007, Measuring and Fostering the Progress of Societies. Available from: http://www.oecd.org/site/0,3407,en_21571361_31938349_1_1_1_1_1,00.html

[21] Heidegger M. The Question Concerning Technology, and Other Essays. 6th ed. New York: Harper & Row; 1977

[22] Winograd T, Flores F. Understanding Computers and Cognition: A New Foundation for Design Norwood. New Jersey: Ablex Publishing Corporation; 1986

[23] Fry T. Becoming Human by Design. Oxford, UK: Berg Publishers; 2012

[24] Partnership on Artificial Intelligence to Benefit People and Society to Formulate Best Practices on Artificial Intelligence Technologies. 2016. Available from: https://www.partnershiponai.org/about/

[25] IEEE. Ethically Aligned Design, A Vision for Prioritizing Human Well-being with Autonomous and Intelligent Systems. First Edition (EAD1e) i. 2018. Available from: https://ethicsinaction.ieee.org/

[26] Cinoğlu H, Arıkan Y. Self, identity and identity formation: From the perspectives of three major theories. International Journal of Human Sciences. 2012;**9**(2):1114-1131

[27] Hewitt JP. Dilemmas of the American Self. Philadelphia: Temple University Press; 1989

[28] Ryff CD, Singer BH. Best news yet on the six-factor model of well-being. Social Science Research. 2006;**35**(4):1103-1119. DOI: 10.1016/j.ssresearch.2006.01.002

[29] OECD. Better Life Initiative – Measuring Progress. 2013. Available from: http://www.oecd.org/sdd/OECD-stat-work-2013-well-being.pdf

[30] Blazer DG. Self-efficacy and depression in late life: A primary prevention proposal. Aging & Mental Health. 2002;**6**(4):315-324

[31] WHO. 10 Facts On Ageing & Health. 2017. Available from: https://www.who.int/features/factfiles/ageing/en/

[32] Rowe JW, Kahn RL. Successful Aging. New York: Pantheon Books; 1988

[33] Frijters P, Beatton T. The mystery of the U-shaped relationship between happiness and age. Journal of Economic Behavior & Organization. 2012;**82**(2-3):525-542. Available from: https://www.sciencedirect.com/science/article/pii/S0167268112000601

[34] Population.sg Team. Smart Ageing Strategies in Japan, Taiwan and Singapore. 2018. Available from: https://www.population.sg/articles/smart-ageing-strategies-in-japan-taiwan-and-singapore

[35] Fuller R. Towards a general theory of driver behaviour. Accident; Analysis and Prevention. 2005;**37**(3):461-472

[36] DaCoTA. Cost-benefit analysis, Deliverable 4.8d of the EC FP7 project DaCoTA, Brussels. 2012

[37] AAA Foundation for Traffic Safety. Relationship Between Driving Habits and Health-Related Quality of Life: AAA LongROAD Study. 2018. Available from: https://aaafoundation.org/relationship-between-driving-habits-and-health-related-quality-of-life-aaa-longroad-study/

[38] Hjorthol R, Levin L, Sirén A. Mobility in different generations of older persons: The development of daily travel in different cohorts in Denmark, Norway and Sweden. Journal of Transport Geography. 2010;**18**(5):624-633

[39] Szenamo Project. Scenarios of the future mobility of elderly people. Final Report. 2010

[40] Van Beek P et al. Senior Cosmopolitan or Modern Traditional: Keep moving. The mobility behaviour of the future older people in three scenario's [In Dutch: Senior Cosmopoliet of Modern Traditioneel: Keep moving Het mobiliteitsgedrag van toekomstige ouderen in drie scenario's]. 2010

[41] Aigner-Breuss E et al. Mobilitätsszenarienkatalog – Projekt MOTION 55+. Vienna, Austria: Kuratorium für Verkehrssicherheit; 2010

[42] Beaudoux M, Deleu H. Mobility for an Ageing Population. Paris, France: Veolia Mobility Lab; 2010

[43] Goal Project. Deliverable D2.1—Profiles of Older People. Growing Older, Staying Mobile: Transport Needs for an Ageing Society. 2013. Available from: http://www.goal-project.eu/images/reports/d2-1_goal_final_20120725.pdf

[44] Langford J, Koppel S. Epidemiology of older driver crashes—Identifying older driver risk factors and exposure patterns. Transportation Research Part F: Traffic Psychology and Behaviour. 2006;**9**(5):309-321

[45] Vichitvanichphong S, Talaei-Khoei A, Kerr D, Ghapanchi A. What does happen to our driving when we get older? Transport Reviews. 2015;**35**(1):56-81

[46] Arai A, Arai Y. Self-assessed driving behaviors associated with age among middle-aged and older adults in Japan. Archives of Gerontology and Geriatrics. 2015;**60**(1):39-44. ISSN: 1872-6976

[47] Polders E, Brijs T, Vlahogianni E, Papadimitriou E, Yannis G, Leopold F, et al. Eldersafe: Risks and countermeasures for road traffic of elderly in Europe. Final report. N MOVE/C4/2014-244. 2016. Available from: https://ec.europa.eu/transport/road_safety/sites/roadsafety/files/pdf/studies/eldersafe_final_report.pdf

[48] Dickerson AE, Molnar LJ, Eby DW, Adler H, Be'dard M, Berg-Weger M, et al. Transportation and aging: A research agenda for advancing safe mobility. The Gerontologist. 2007;**47**:578-590

[49] Charlton J, Koppel S, Odell M, Devlin A, Langford J, O'Hare M, et al. Influence of Chronic Illness on Crash Involvement of Motor Vehicle Drivers: 2nd Edition (No. 300). Victoria, Australia: MUARC (Monash University Accident Research Centre); 2010

[50] Marshall SC. The role of reduced fitness to drive due to medical impairments in explaining crashes involving older drivers. Traffic Injury Prevention. 2008;**9**(4):291-298. DOI: 10.1080/15389580801895244

[51] Thrun S. Toward robotic cars. Communications of the

ACM. 2010;**53**(4):99-106. DOI: 10.1145/1721654.1721679

[52] SAE International. J3016, International Taxonomy and Definitions for Terms Related to Driving Automation Systems for On-Road Motor Vehicles (J3016:Sept 2016)

[53] Gasser TM, Westhoff D. BASt-Study: Definitions of Automation and Legal Issues in Germany. 2012

[54] Rouse, M. Self-driving Car (Autonomous Car or Driverless Car). 2019. Available from: https://searchenterpriseai.techtarget.com/definition/driverless-car

[55] Vienna Convention on Road Traffic. Global Automotive Regulation Page Related to Automated Driving. 2017. Available from: https://globalautoregs.com/rules/157-1968-vienna-convention-on-road-traffic

[56] BMVI. Ethics Commission—Appointed by the Federal Minister of Transport and Digital Infrastructure—Connected and Automated Driving (2017)—BMVI.DE

[57] Surakitbanharn CA. Preliminary Ethical, Legal and Social Implications of Connected and Autonomous Transportation Vehicles (CATV). 2018

[58] Nordhoff S, de Winter J, Kyriakidis M, van Arem B, Happee R. Acceptance of driverless vehicles: Results from a large cross-national questionnaire study. Journal of Advanced Transportation. 2018;**2018**:Article ID 5382192, 22 p. Available from: 10.1155/2018/5382192

[59] Schoettle B, Sivak M. A Survey of Public Opinion about Autonomous and Self-driving Vehicles in the U.., the U.K., and Australia (Technical Report No. UMTRI-2014-21). 2014. Available from: http://deepblue.lib.umich.edu/bitstream/handle/2027.42/108384/103024.pdf?sequence=1&isAllowed=y

[60] Pruitt J, Grudin J. Personas: Practice and theory. In: Proceedings of the 2003 Conference on Designing for User Experiences (DUX '03). New York, NY, USA: ACM; 2003. pp. 1-15. DOI: 10.1145/997078.997089

[61] Carroll JM. Scenario-based Design: Envisioning Work and Technology in System Development. New York: John Wiley and Sons; 1995

[62] Bødker S, Burr J. The design collaboratorium. A place for usability design. ACM Transactions on Computer Human Interaction. 2002;**9**(2):152-169

[63] Yin S. Smoke, fire and human evolution. The New York Times. 2016. ISSN 0362-4331. Available from: https://www.nytimes.com/2016/08/09/science/fire-smoke-evolution-tuberculosis.html

[64] Kahn CH. The Art and Thought of Heraclitus. Cambridge: Cambridge University Press; 1979

[65] Salganik MJ. Bit by Bit: Social Research in the Digital Age. Open review ed. NJ: Princeton University Press; 2017

[66] Turner N, Donoghue O, Kenny RA. Wellbeing and Health in Ireland's over 50s 2009-2016. 2018. Available from: https://tilda.tcd.ie/publications/reports/pdf/w4-key-findings-report/TILDA-Wave4-Key-Findings-report.pdf

Chapter 3

Social Media, Ethics and the Privacy Paradox

Nadine Barrett-Maitland and Jenice Lynch

Abstract

Today's information/digital age offers widespread use of social media. The use of social media is ubiquitous and cuts across all age groups, social classes and cultures. However, the increased use of these media is accompanied by privacy issues and ethical concerns. These privacy issues can have far-reaching professional, personal and security implications. Ultimate privacy in the social media domain is very difficult because these media are designed for sharing information. Participating in social media requires persons to ignore some personal, privacy constraints resulting in some vulnerability. The weak individual privacy safeguards in this space have resulted in unethical and undesirable behaviors resulting in privacy and security breaches, especially for the most vulnerable group of users. An exploratory study was conducted to examine social media usage and the implications for personal privacy. We investigated how some of the requirements for participating in social media and how unethical use of social media can impact users' privacy. Results indicate that if users of these networks pay attention to privacy settings and the type of information shared and adhere to universal, fundamental, moral values such as mutual respect and kindness, many privacy and unethical issues can be avoided.

Keywords: privacy, ethics, social media

1. Introduction

The use of social media is growing at a rapid pace and the twenty-first century could be described as the "boom" period for social networking. According to reports provided by Smart Insights, as at February 2019 there were over 3.484 billion social media users. The Smart Insight report indicates that the number of social media users is growing by 9% annually and this trend is estimated to continue. Presently the number of social media users represents 45% of the global population [1]. The heaviest users of social media are "digital natives"; the group of persons who were born or who have grown up in the digital era and are intimate with the various technologies and systems, and the "Millennial Generation"; those who became adults at the turn of the twenty-first century. These groups of users utilize social media platforms for just about anything ranging from marketing, news acquisition, teaching, health care, civic engagement, and politicking to social engagement.

The unethical use of social media has resulted in the breach of individual privacy and impacts both physical and information security. Reports in 2019 [1], reveal that persons between the ages 8 and 11 years spend an average 13.5 hours weekly online and 18% of this age group are actively engaged on social media. Those between ages 12 and 15 spend on average 20.5 hours online and 69% of this group are active social media users. While children and teenagers represent the largest Internet user groups, for the most part they do not know how to protect their personal information on the Web and are the most vulnerable to cyber-crimes related to breaches of information privacy [2, 3].

In today's IT-configured society data is one of, if not the most, valuable asset for most businesses/organizations. Organizations and governments collect information via several means including invisible data gathering, marketing platforms and search engines such as Google [4]. Information can be attained from several sources, which can be fused using technology to develop complete profiles of individuals. The information on social media is very accessible and can be of great value to individuals and organizations for reasons such as marketing, etc.; hence, data is retained by most companies for future use.

2. Privacy

Privacy or the right to enjoy freedom from unauthorized intrusion is the negative right of all human beings. Privacy is defined as the right to be left alone, to be free from secret surveillance, or unwanted disclosure of personal data or information by government, corporation, or individual (dictionary.com). In this chapter we will define privacy loosely, as the right to control access to personal information. Supporters of privacy posit that it is a necessity for human dignity and individuality and a key element in the quest for happiness. According to Baase [5] in the book titled "A Gift of Fire: Social, Legal and Ethical Issues for Computing and the Internet," privacy is the ability to control information about one' s self as well as the freedom from surveillance from being followed, tracked, watched, and being eavesdropped on. In this regard, ignoring privacy rights often leads to encroachment on natural rights.

Privacy, or even the thought that one has this right, leads to peace of mind and can provide an environment of solitude. This solitude can allow people to breathe freely in a space that is free from interference and intrusion. According to Richards and Solove [6], Legal scholar William Prosser argued that privacy cases can be classified into four related "torts," namely:

1. Intrusion—this can be viewed as encroachment (physical or otherwise) on ones liberties/solitude in a highly offensive way.

2. Privacy facts—making public, private information about someone that is of no "legitimate concern" to anyone.

3. False light—making public false and "highly offensive" information about others.

4. Appropriation—stealing someone's identity (name, likeness) to gain advantage without the permission of the individual.

Technology, the digital age, the Internet and social media have redefined privacy however as surveillance is no longer limited to a certain pre-defined space and location. An understanding of the problems and dangers of privacy in the digital space is therefore the first step to privacy control. While there can be clear distinctions between informational privacy and physical privacy, as pointed out earlier, intrusion can be both physical and otherwise.

This chapter will focus on informational privacy which is the ability to control access to personal information. We examine privacy issues in the social media context focusing primarily on personal information and the ability to control external influences. We suggest that breach of informational privacy can impact: solitude (the right to be left alone), intimacy (the right not to be monitored), and anonymity (the right to have no public personal identity and by extension physical privacy impacted). The right to control access to facts or personal information in our view is a natural, inalienable right and everyone should have control over who see their personal information and how it is disseminated.

In May 2019 the General Data Protection Regulation (GDPR) clearly outlined that it is unlawful to process personal data without the consent of the individual (subject). It is a legal requirement under the GDPR that privacy notices be given to individuals that outline how their personal data will be processed and the conditions that must be met that make the consent valid. These are:

1. "Freely given—an individual must be given a genuine choice when providing consent and it should generally be unbundled from other terms and conditions (e.g., access to a service should not be conditional upon consent being given)."

2. "Specific and informed—this means that data subjects should be provided with information as to the identity of the controller(s), the specific purposes, types of processing, as well as being informed of their right to withdraw consent at any time."

3. "Explicit and unambiguous—the data subject must clearly express their consent (e.g., by actively ticking a box which confirms they are giving consent—pre-ticked boxes are insufficient)."

4. "Under 13s—children under the age of 13 cannot provide consent and it is therefore necessary to obtain consent from their parents."

Arguments can be made that privacy is a cultural, universal necessity for harmonious relationships among human beings and creates the boundaries for engagement and disengagement. Privacy can also be viewed as instrumental good because it is a requirement for the development of certain kinds of human relationships, intimacy and trust [7]. However, achieving privacy is much more difficult in light of constant surveillance and the inability to determine the levels of interaction with various publics [7]. Some critics argue that privacy provides protection against anti-social behaviors such as trickery, disinformation and fraud, and is thought to be a universal right [5]. However, privacy can also be viewed as relative as privacy rules may differ based on several factors such as "climate, religion, technological advancement and political arrangements" [8, 9]. The need for privacy is an objective reality though it can be viewed as "culturally rational" where the need for personal privacy is viewed as relative based on culture. One example is the push by the government, businesses and Singaporeans to make Singapore a smart nation. According to GovTech 2018 reports there is a push by the government in Singapore to harness the data "new gold" to develop systems that can make life easier for its people. The [10] report points out that Singapore is using sensors robots Smart Water Assessment Network (SWAN) to monitor water quality in its reservoirs, seeking to build smart health system and to build a smart transportation system to name a few. In this example privacy can be describe as "culturally rational" and the rules in general could differ based on technological advancement and political arrangements.

In today's networked society it is naïve and ill-conceived to think that privacy is over-rated and there is no need to be concerned about privacy if you have done nothing wrong [5]. The effects of information flow can be complex and may not be simply about protection for people who have something to hide. Inaccurate information flow can have adverse long-term implications for individuals and companies. Consider a scenario where someone's computer or tablet is stolen. The perpetrator uses identification information stored on the device to access their social media page which could lead to access to their contacts, friends and friends of their "friends" then participate in illegal activities and engage in anti-social activities such as hacking, spreading viruses, fraud and identity theft. The victim is now in danger of being accused of criminal intentions, or worse. These kinds of situations are possible because of technology and networked systems. Users of social media need to be aware of the risks that are associated with participation.

3. Social media

The concept of social networking pre-dates the Internet and mass communication as people are said to be social creatures who when working in groups can achieve results in a value greater than the sun of its parts [11]. The explosive growth in the use of social media over the past decade has made it one of the most popular Internet services in the world, providing new avenues to "see and be seen" [12, 13]. The use of social media has changed the communication landscape resulting in changes in ethical norms and behavior. The unprecedented level of growth in usage has resulted in the reduction in the use of other media and changes in areas including civic and political engagement, privacy and safety [14]. Alexa, a company that keeps track of traffic on the Web, indicates that as of August, 2019 YouTube, Facebook and Twitter are among the top four (4) most visited sites with only Google, being the most popular search engine, surpassing these social media sites.

Social media sites can be described as online services that allow users to create profiles which are "public, semi-public" or both. Users may create individual profiles and/or become a part of a group of people with whom they may be acquainted offline [15]. They also provide avenues to create virtual friendships. Through these virtual friendships, people may access details about their contacts ranging from personal background information and interests to location. Social networking sites provide various tools to facilitate communication. These include chat rooms, blogs, private messages, public comments, ways of uploading content external to the site and sharing videos and photographs. Social media is therefore drastically changing the way people communicate and form relationships.

Today social media has proven to be one of the most, if not the most effective medium for the dissemination of information to various audiences. The power of this medium is phenomenal and ranges from its ability to overturn governments (e.g., Moldova), to mobilize protests, assist with getting support for humanitarian aid, organize political campaigns, organize groups to delay the passing of legislation (as in the case with the copyright bill in Canada) to making social media billionaires and millionaires [16, 17]. The enabling nature and the structure of the media that social networking offers provide a wide range of opportunities that were nonexistent before technology. Facebook and YouTube marketers and trainers provide two examples. Today people can interact with and learn from people millions of miles away. The global reach of this medium has removed all former pre-defined boundaries including geographical, social and any other that existed previously.

Technological advancements such as Web 2.0 and Web 4.0 which provide the framework for collaboration, have given new meaning to life from various perspectives: political, institutional and social.

4. Privacy and social media

Social medial and the information/digital era have "redefined" privacy. In today's Information Technology—configured societies, where there is continuous monitoring, privacy has taken on a new meaning. Technologies such as closed-circuit cameras (CCTV) are prevalent in public spaces or in some private spaces including our work and home [7, 18]. Personal computers and devices such as our smart phones enabled with Global Positioning System (GPS), Geo locations and Geo maps connected to these devices make privacy as we know it, a thing of the past. Recent reports indicate that some of the largest companies such as Amazon, Microsoft and Facebook as well as various government agencies are collecting information without consent and storing it in databases for future use. It is almost impossible to say privacy exists in this digital world (@nowthisnews).

The open nature of the social networking sites and the avenues they provide for sharing information in a "public or semi-public" space create privacy concerns by their very construct. Information that is inappropriate for some audiences are many times inadvertently made visible to groups other than those intended and can sometimes result in future negative outcomes. One such example is a well-known case recorded in an article entitled "The Web Means the End of Forgetting" that involved a young woman who was denied her college license because of backlash from photographs posted on social media in her private engagement.

Technology has reduced the gap between professional and personal spaces and often results in information exposure to the wrong audience [19]. The reduction in the separation of professional and personal spaces can affect image management especially in a professional setting resulting in the erosion of traditional professional image and impression management. Determining the secondary use of personal information and those who have access to this information should be the prerogative of the individual or group to whom the information belongs. However, engaging in social media activities has removed this control.

Privacy on social networking sites (SNSs) is heavily dependent on the users of these networks because sharing information is the primary way of participating in social communities. Privacy in SNSs is "multifaceted." Users of these platforms are responsible for protecting their information from third-party data collection and managing their personal profiles. However, participants are usually more willing to give personal and more private information in SNSs than anywhere else on the Internet. This can be attributed to the feeling of community, comfort and family that these media provide for the most part. Privacy controls are not the priority of social networking site designers and only a small number of the young adolescent users change the default privacy settings of their accounts [20, 21]. This opens the door for breaches especially among the most vulnerable user groups, namely young children, teenagers and the elderly. The nature of social networking sites such as Facebook and Twitter and other social media platforms cause users to re-evaluate and often change their personal privacy standards in order to participate in these social networked communities [13].

While there are tremendous benefits that can be derived from the effective use of social media there are some unavoidable risks that are involved in its use. Much attention should therefore be given to what is shared in these forums. Social platforms such as Facebook, Twitter and YouTube are said to be the most effective media to communicate to Generation Y's (Gen Y's), as teens and young adults are the largest user groups on these platforms [22]. However, according to Bolton et al. [22] Gen Y's use of social media, if left unabated and unmonitored will have long-term implications for privacy and engagement in civic activities as this continuous use is resulting in changes in behavior and social norms as well as increased levels of cyber-crime.

Today social networks are becoming the platform of choice for hackers and other perpetrators of antisocial behavior. These media offer large volumes of data/information ranging from an individual's date of birth, place of residence, place of work/business, to information about family and other personal activities. In many cases users unintentionally disclose information that can be both dangerous and inappropriate. Information regarding activities on social media can have far reaching negative implications for one's future. A few examples of situations which can, and have been affected are employment, visa acquisition, and college acceptance. Indiscriminate participation has also resulted in situations such identity theft and bank fraud just to list a few. Protecting privacy in today's networked society can be a great challenge. The digital revolution has indeed distorted our views of privacy, however, there should be clear distinctions between what should be seen by the general public and what should be limited to a selected group. One school of thought is that the only way to have privacy today is not to share information in these networked communities. However, achieving privacy and control over information flows and disclosure in networked communities is an ongoing process in an environment where contexts change quickly and are sometimes blurred. This requires intentional construction of systems that are designed to mitigate privacy issues [13].

5. Ethics and social media

Ethics can be loosely defined as "the right thing to do" or it can be described as the moral philosophy of an individual or group and usually reflects what the individual or group views as good or bad. It is how they classify particular situations by categorizing them as right or wrong. Ethics can also be used to refer to any classification or philosophy of moral values or principles that guides the actions of an individual or group [23]. Ethical values are intended to be guiding principles that if followed, could yield harmonious results and relationships. They seek to give answers to questions such as "How should I be living? How do I achieve the things that are deemed important such as knowledge and happiness or the acquisition of attractive things?" If one chooses happiness, the next question that needs to be answered is "Whose happiness should it be; my own happiness or the happiness of others?" In the domain of social media, some of the ethical questions that must be contemplated and ultimately answered are [24]:

- Can this post be regarded as oversharing?
- Has the information in this post been distorted in anyway?
- What impact will this post have on others?

As previously mentioned, users within the ages 8–15 represent one of the largest social media user groups. These young persons within the 8–15 age range are still learning how to interact with the people around them and are deciding on the moral values that they will embrace. These moral values will help to dictate how they will interact with the world around them. The ethical values that guide our interactions are usually formulated from some moral principle taught to us by someone or a group of individuals including parents, guardians, religious groups, and teachers just to name a few. Many of the Gen Y's/"Digital Babies" are "newbies" yet are required to determine for themselves the level of responsibility they will display when using the varying social media platforms. This includes considering the impact a post will have on their lives and/or the lives of other persons. They must also understand that when they join a social media network, they are joining a community in which certain behavior must be exhibited. Such responsibility requires a much greater level of maturity than can be expected from them at that age.

It is not uncommon for individuals to post even the smallest details of their lives from the moment they wake up to when they go to bed. They will openly share their location, what they eat at every meal or details about activities typically considered private and personal. They will also share likes and dislikes, thoughts and emotional states and for the most part this has become an accepted norm. Often times however, these shares do not only contain information about the person sharing but information about others as well. Many times, these details are shared on several social media platforms as individuals attempt to ensure that all persons within their social circle are kept updated on their activities. With this openness of sharing risks and challenges arise that are often not considered but can have serious impacts. The speed and scale with which social media creates information and makes it available—almost instantaneously—on a global scale, added to the fact that once something is posted there is really no way of truly removing it, should prompt individuals to think of the possible impact a post can have. Unfortunately, more often than not, posts are made without any thought of the far-reaching impact they can have on the lives of the person posting or others that may be implicated by the post.

6. Why do people share?

According to Berger and Milkman [25] there are five (5) main reasons why users are compelled to share content online, whether it is every detail or what they deem as highlights of their lives. These are:

- cause related
- personal connection to content
- to feel more involved in the world
- to define who they are
- to inform and entertain

People generally share because they believe that what they are sharing is important. It is hoped that the shared content will be deemed important to others which will ultimately result in more shares, likes and followers.

Figure 1 below sums up the findings of Berger and Milkman [25] which shows that the main reason people feel the need to share content on the varying social media platform is that the content relates to what is deemed as worthy cause. 84% of respondents highlighted this as the primary motivation for sharing. Seventy-eight percent said that they share because they feel a personal connection to the content while 69 and 68%, respectively said the content either made them feel more

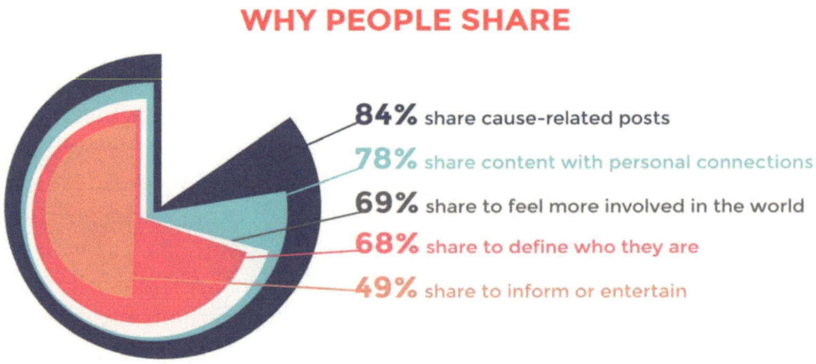

Figure 1.
Why people share source: Global Social Media Research. thesocialmediahat.com [26].

involved with the world or helped them to define who they were. Forty-nine percent share because of the entertainment or information value of the content. A more in depth look at each reason for sharing follows.

7. Content related to a cause

Social media has provided a platform for people to share their thoughts and express concerns with others for what they regard as a worthy cause. Cause related posts are dependent on the interest of the individual. Some persons might share posts related to causes and issues happening in society. In one example, the parents of a baby with an aggressive form of leukemia, who having been told that their child had only 3 months to live unless a suitable donor for a blood stem cell transplant could be found, made an appeal on social media. The appeal was quickly shared and a suitable donor was soon found. While that was for a good cause, many view social media merely as platforms for freedom of speech because anyone can post any content one creates. People think the expression of their thoughts on social media regarding any topic is permissible. The problem with this is that the content may not be accepted by law or it could violate the rights of someone thus giving rise to ethical questions.

8. Content with a personal connection

When social media users feel a personal connection to their content, they are more inclined to share the content within their social circles. This is true of information regarding family and personal activities. Content created by users also invokes a deep feeling of connection as it allows the users to tell their stories and it is natural to want the world or at least friends to know of the achievement. This natural need to share content is not new as humans have been doing this in some form or the other, starting with oral history to the media of the day; social media. Sharing the self-created content gives the user the opportunity of satisfying some fundamental needs of humans to be heard, to matter, to be understood and emancipated. The problem with this however is that in an effort to gratify the fundamental needs, borders are crossed because the content may not be sharable (can this content be shared within the share network?), it may not be share-worthy (who is the audience that would appreciate this content?) or it may be out of context (does the content fit the situation?).

9. Content that makes them feel more involved in the world

One of the driving factors that pushes users to share content is the need to feel more in tune with the world around them. This desire is many times fueled by jealousy. Many social media users are jealous when their friends' content gets more attention than their own and so there is a lot of pressure to maintain one's persona in social circles, even when the information is unrealistic, as long as it gets as much attention as possible. Everything has to be perfect. In the case of a photo, for example, there is lighting, camera angle and background to consider. This need for perfection puts a tremendous amount of pressure on individuals to ensure that posted content is "liked" by friends. They often give very little thought to the amount of their friend's work that may have gone on behind the scenes to achieve that perfect social post.

Social media platforms have provided everyone with a forum to express views, but, as a whole, conversations are more polarized, tribal and hostile. With Facebook for instance, there has been a huge uptick in fake news, altered images, dangerous health claims and cures, and the proliferation of anti-science information. This is very distressing and disturbing because people are too willing to share and to believe without doing their due diligence and fact-checking first.

10. Content that defines who they are

Establishing one's individuality in society can be challenging for some persons because not everyone wants to fit in. Some individuals will do all they can to stand out and be noticed. Social media provides the avenue for exposure and many individuals will seek to leverage the media to stand out of the crowd and not just be a fish in the school. Today many young people are currently being brought up in a culture that defines people by their presence on social media where in previous generations, persons were taught to define themselves by their career choices. These lessons would start from childhood by asking children what they wanted to be when they grew up and then rewarding them based on the answers they give [27]. In today's digital era, however, social media postings and the number of "likes" or "dislikes" they attract, signal what is appealing to others. Therefore, post that are similar to those that receive a large number of likes but which are largely unrealistic are usually made for self-gratification.

11. Content that informs and entertains

The acquisition of knowledge and skills is a vital part of human survival and social media has made this process much easier. It is not uncommon to hear persons realizing that they need a particular knowledge set that they do not possess say "I need to lean to do this. I'll just YouTube it." Learning and adapting to change in as short as possible time is vital in today's society and social media coupled with the Internet put it all at the finger tips. Entertainment has the ability to bring people together and is a good way for people to bond. It provides a diversion from the demands of life and fills leisure time with amusement. Social media is an outlet for fun, pleasurable and enjoyable activities that are so vital to human survival [28]. It is now common place to see persons watching a video, viewing images and reading text that is amusing on any of the available social media platforms. Quite often these videos, images and texts can be both informative and entertaining, but there can be problems however as at times they can cross ethical lines that can lead to conflict.

12. Ethical challenges with social media use

The use of modern-day technology has brought several benefits. Social media is no different and chief amongst its benefit is the ability to stay connected easily and quickly as well as build relationships with people with similar interests. As with all technology, there are several challenges that can make the use of social media off putting and unpleasant. Some of these challenges appear to be minor but they can have far reaching effects into the lives of the users of social media and it is therefore advised that care be taken to minimize the challenges associated with the use of social media [29].

A major challenge with the use of social media is oversharing because when persons share on social media, they tend to share as much as is possible which is often times too much [24]. When persons are out and about doing exciting things, it is natural to want to share this with the world as many users will post a few times a day when they head to lunch, visit a museum, go out to dinner or other places of interest [30]. While this all seems relatively harmless, by using location-based services which pinpoint users with surprising accuracy and in real time, users place themselves in danger of laying out a pattern of movement that can be easily traced. While this seems more like a security or privacy issue it stems from an ethical dilemma—"Am I sharing too much?" Oversharing can also lead to damage of user's reputation especially if the intent is to leverage the platform for business [24]. Photos of drunken behavior, drug use, partying or other inappropriate content can change how you are viewed by others.

Another ethical challenge users of social media often encounter is that they have no way of authenticating content before sharing, which becomes problematic when the content paints people or establishments negatively. Often times content is shared with them by friends, family and colleagues. The unauthenticated content is then reshared without any thought but sometimes this content may have been maliciously altered so the user unknowingly participates in maligning others. Even if the content is not altered the fact that the content paints someone or something in a bad light should send off warning bells as to whether or not it is right to share the content which is the underlying principle of ethical behavior.

13. Conflicting views

Some of the challenges experienced by social media posts are a result of a lack of understanding and sometimes a lack of respect for the varying ethical and moral standpoints of the people involved. We have established that it is typical for persons to post to social media sites without any thought as to how it can affect other persons, but many times these posts are a cause of conflict because of a difference of opinion that may exist and the effect the post may have. Each individual will have his or her own ethical values and if they differ then this can result in conflict [31]. When an executive of a British company made an Instagram post with some racial connotations before boarding a plane to South Africa it started a frenzy that resulted in the executive's immediate dismissal. Although the executive said it was a joke and there was no prejudice intended, this difference in views as to the implications of the post, resulted in an out of work executive and a company scrambling to maintain its public image.

14. Impact on personal development

In this age of sharing, many young persons spend a vast amount of time on social media checking the activities of their "friends" as well as posting on their own activities so their "friends" are aware of what they are up to. Apart from interfering with their academic progress, time spent on these posts at can have long term repercussions. An example is provided by a student of a prominent university who posted pictures of herself having a good time at parties while in school. She was denied employment because of some of her social media posts. While the ethical challenge here is the question of the employee's right to privacy and whether the individual's social media profile should affect their ability to fulfill their responsibilities as an employee, the impact on the individual's long term personal growth is clear.

15. Conclusion

In today's information age, one's digital footprint can make or break someone; it can be the deciding factor on whether or not one achieves one's life-long ambitions. Unethical behavior and interactions on social media can have far reaching implications both professionally and socially. Posting on the Internet means the "end of forgetting," therefore, responsible use of this medium is critical. The unethical use of social media has implications for privacy and can result in security breaches both physically and virtually. The use of social media can also result in the loss of privacy as many users are required to provide information that they would not divulge otherwise. Social media use can reveal information that can result in privacy breaches if not managed properly by users. Therefore, educating users of the risks and dangers of the exposure of sensitive information in this space, and encouraging vigilance in the protection of individual privacy on these platforms is paramount. This could result in the reduction of unethical and irresponsible use of these media and facilitate a more secure social environment. The use of social media should be governed by moral and ethical principles that can be applied universally and result in harmonious relationships regardless of race, culture, religious persuasion and social status.

Analysis of the literature and the findings of this research suggest achieving acceptable levels of privacy is very difficult in a networked system and will require much effort on the part of individuals. The largest user groups of social media are unaware of the processes that are required to reduce the level of vulnerability of their personal data. Therefore, educating users of the risk of participating in social media is the social responsibility of these social network platforms. Adapting universally ethical behaviors can mitigate the rise in the number of privacy breaches in the social networking space. This recommendation coincides with philosopher Immanuel Kant's assertion that, the Biblical principle which states "Do unto others as you have them do unto you" can be applied universally and should guide human interactions [5]. This principle, if adhered to by users of social media and owners of these platforms could raise the awareness of unsuspecting users, reduce unethical interactions and undesirable incidents that could negatively affect privacy, and by extension security in this domain.

Author details

Nadine Barrett-Maitland and Jenice Lynch
University of Technology, Jamaica, West Indies

*Address all correspondence to: nadinemland@yahoo.com

References

[1] Chaffey D. Global Social Media Research. Smart Insights. 2019. Retrieved from: https://www.smartinsights.com/social-media-marketing/social-media-strategy/new-global-social-media-research/

[2] SmartSocial. Teen Social Media Statistics (What Parents Need to Know). 2019. Retrieved from: https://smartsocial.com/social-media-statistics/

[3] Wisniewski P, Jia H, Xu H, Rosson MB, Carroll JM. Preventative vs. reactive: How parental mediation influences teens' social media privacy behaviors. In: Proceedings of the 18th ACM Conference on Computer Supported Cooperative Work and Social Computing; ACM; 2015. pp. 302-316

[4] Chai S, Bagchi-Sen S, Morrell C, Rao HR, Upadhyaya SJ. Internet and online information privacy: An exploratory study of preteens and early teens. IEEE Transactions on Professional Communication. 2009;**52**(2):167-182

[5] Baase S. A Gift of Fire. Upper Saddle River, New Jersey: Pearson Education Limited (Prentice Hall); 2012

[6] Richards NM, Solove DJ. Prosser's privacy law: A mixed legacy. California Law Review. 2010;**98**:1887

[7] Johnson DG. Computer ethics. In: The Blackwell Guide to the Philosophy of Computing and Information. Upper Saddle River, New Jersey: Pearson Education (Prentice Hall); 2004. pp. 65-75

[8] Cohen JE. What privacy is for. Harvard Law Review. 2012;**126**:1904

[9] Moore AD. Toward informational privacy rights. San Diego Law Review. 2007;**44**:809

[10] GOVTECH. Singapore. 2019. Retrieved from: https://www.tech.gov.sg/products-and-services/smart-nation-sensor-platform/

[11] Weaver AC, Morrison BB. Social networking. Computer. 2008;**41**(2):97-100

[12] Boulianne S. Social media use and participation: A meta-analysis of current research. Information, Communication and Society. 2015;**18**(5):524-538

[13] Marwick AE, Boyd D. Networked privacy: How teenagers negotiate context in social media. New Media & Society. 2014;**16**(7):1051-1067

[14] McCay-Peet L, Quan-Haase A. What is social media and what questions can social media research help us answer. In: The SAGE Handbook of Social Media Research Methods. Thousand Oaks, CA: SAGE Publishers; 2017. pp. 13-26

[15] Gil de Zúñiga H, Jung N, Valenzuela S. Social media use for news and individuals' social capital, civic engagement and political participation. Journal of Computer-Mediated Communication. 2012;**17**(3):319-336

[16] Ems L. Twitter's place in the tussle: How old power struggles play out on a new stage. Media, Culture and Society. 2014;**36**(5):720-731

[17] Haggart B. Fair copyright for Canada: Lessons for online social movements from the first Canadian Facebook uprising. Canadian Journal of Political Science (Revue canadienne de science politique). 2013;**46**(4):841-861

[18] Andrews LB. I Know Who You are and I Saw What You Did: Social Networks and the Death of Privacy. Simon and Schuster, Free Press; 2012

[19] Echaiz J, Ardenghi JR. Security and online social networks. In: XV

Congreso Argentino de Ciencias de la Computación. 2009

[20] Barrett-Maitland N, Barclay C, Osei-Bryson KM. Security in social networking services: A value-focused thinking exploration in understanding users' privacy and security concerns. Information Technology for Development. 2016;**22**(3):464-486

[21] Van Der Velden M, El Emam K. "Not all my friends need to know": A qualitative study of teenage patients, privacy, and social media. Journal of the American Medical Informatics Association. 2013;**20**(1):16-24

[22] Bolton RN, Parasuraman A, Hoefnagels A, Migchels N, Kabadayi S, Gruber T, et al. Understanding Generation Y and their use of social media: A review and research agenda. Journal of Service Management. 2013;**24**(3):245-267

[23] Cohn C. Social Media Ethics and Etiquette. CompuKol Communication LLC. 20 March 2010. Retrieved from: https://www.compukol.com/social-media-ethics-and-etiquette/

[24] Nates C. The Dangers of Oversharing of Social Media. Pure Moderation. 2018. Retrieved from: https://www.puremoderation.com/single-post/The-Dangers-of-Oversharing-on-Social-Media

[25] Berger J, Milkman K. What makes online content go viral. Journal of Marketing Research. 2011;**49**(2):192-205

[26] The Social Media Hat. How to Find Amazing Content for Your Social Media Calendar (And Save Yourself Tons of Work). 29 August 2016. Retrieved from: https://www.thesocialmediahat.com/blog/how-to-find-amazing-content-for-your-social-media-calendar-and-save-yourself-tons-of-work/

[27] People First. Does what you do define who you are. 15 September 2012. Retrieved from: https://blog.peoplefirstps.com/connect2lead/what-you-do-define-you

[28] Dreyfus E. Does what you do define who you are. Psychologically Speaking. 2010. Retrieved from: https://www.edwarddreyfusbooks.com/psychologically-speaking/does-what-you-do-define-who-you-are/

[29] Business Ethics Briefing. The Ethical Challenges and Opportunities of Social Media Use. (Issue 66). 2019. Retrieved from: https://www.ibe.org.uk/userassets/briefings/ibe_social_media_briefing.pdf

[30] Staff Writer. The consequences of oversharing on social networks. Reputation Defender. 2018. Retrieved from: https://www.reputationdefender.com/blog/social-media/consequences-oversharing-social-networks

[31] Business Ethics Briefing. The Ethical Challenges of Social Media. (Issue 22). 2011. Retrieved from: https://www.ibe.org.uk/userassets/briefings/ibe_briefing_22_the_ethical_challenges_of_social_media.pdf

Section 2

Technical Security and Privacy Issues

Chapter 4

Security and Privacy in Three States of Information

Ebru Celikel Cankaya

Abstract

In regard to computational context, information can be in either of three states at a time: in transit, in process, or in storage. When the security and privacy of information is of concern, each of these states should be addressed exclusively, i.e., network security, computer security, and database/cloud security, respectively. This chapter first introduces the three states of information and then addresses the security as well as privacy issues that relate to each state. It provides practical examples for each state discussed, introduces corresponding security and privacy algorithms for explaining the concepts, and facilitates their implementation whenever needed. Moreover, the security and privacy techniques pertaining to the three states of information are combined together to offer a more comprehensive and realistic consideration of everyday security practices.

Keywords: information in transit, information in process, information in storage, network security, computer security, database security, cloud security

1. Introduction

The world is living in an era of information outburst thanks to rapid speed of technological developments in computational domain. Many factors contribute to the immense amount of data available online: Mobile devices are becoming more accessible to everyone with their decreasing cost, underlying network technologies are facilitating data transfer by providing faster and more reliable communication, data processing methods (such as editing, compressing) are helping customize data format, and increasing data storage capacities are turning bulk data storage more effective.

With all technology and convenience so easily reachable, an ordinary user is allured further to handle even more information by either transferring, processing, or storing them. Yet, the issues of security and privacy are often taken for granted, and unfortunately this ignorance may cause a destructive consequence by totally violating the security and privacy of information, as well as its owner.

In this chapter, three fundamental states of information, i.e., information in transit, information in process, and information in storage, are defined first. Then the chapter addresses security and privacy of information in each state by giving examples. It should be noted that as information is the processed form of data, these two terms, i.e., information and data, are used interchangeably throughout the text. The chapter then provides algorithms to achieve privacy and security of information in these three states and how they apply to particular states of information for ensuring security.

The rest of the chapter is organized as follows: In Section 2, three states of information are defined together with examples for each. In Section 3, security mechanisms, i.e., privacy and security algorithms that apply to each state of information, are discussed in detail, and examples are provided. Finally, the chapter is concluded in Section 4.

2. Three states of information

This section identifies three fundamental states that information can be in at a time. It is essential to distinctively identify each state, as corresponding security measures vary for each of these states. To address each separately, the information residing in either of three states at a time in a computational environment are identified. These three states are listed as information in transit, information in process, and information in storage. Information in transit refers to the status where an underlying network (wired or wireless) facilitates the transmission of data from one place (source) to another (destination). Information in process refers to the case when data is processed so that it transforms from source format to destination format. And information in storage refers to the mostly stagnant form of data that resides on a storage media for future reference. As these brief descriptions imply, information in each state has different properties than information in other states. As an example, information in transit is different from information in process and information in storage.

It should also be noted that these states may overlap to form many variations. As an example, states can form variations in pairs, such as information could be processed first and then transferred, or processed first and then stored, or stored information is retrieved for further processing, etc. Or, in a more complex format, each of the three states of information may combine in any order to form a sequence of operations, such as stored information is retrieved from database and is processed to yield new information and this new information is transferred to a remote destination.

The following subsections address each state of information in detail by describing them first and then providing examples for each state. This, later, serves as a basis for explaining security and privacy measures with respect to each state of information that is explained later in Section 3.

2.1 Information in transit

The first state of information is information in transit. This state refers to the situation when information handled is transferred from one place (source) to another place (destination). As depicted in **Figure 1**, in the context of information in transit state, the information residing in source side is transmitted to destination via an underlying network. The underlying network infrastructure could be of various types, such as cable network, wireless network, etc., and does not differentiate

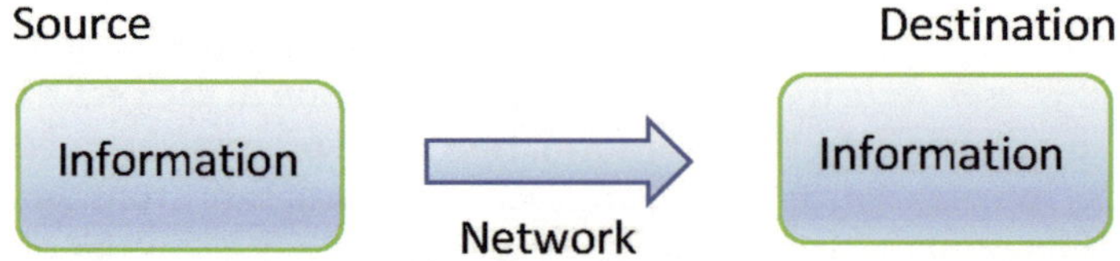

Figure 1.
Information in transit.

the type of data being transferred, as each data piece is processed in the bit level of granularity. This enables the transfer of information in various possible formats, such as plaintext, still image, movie, voice, etc.

Though **Figure 1** illustrates information being transmitted from source to destination, more than usual, the roles change and the source becomes destination and the destination becomes source. This is because of the inherently full duplex property of communication in most of the times.

It should also be noted that if information in transit is considered only in isolation, then it should not change the format of data being transferred. This explains why in **Figure 1** both sender and destination sides label information exactly the same way. And to guarantee this unchanging property of data, integrity mechanisms are inherently built in network systems that facilitate information in transit. If a more complex system is designed and implemented by combining information in transit and information in process states, then labeling of each information entity will have to be modified accordingly. In particular, these entities will label information as $Information_1$ and $Information_2$ on each side so as to indicate the changing content of the information itself. This phenomenon is explained further in Section 2.2 and is depicted later in **Figure 2**.

Sending a memo that has text, pictures, and videos in it as part of a business operation from company headquarters to a branch office is an example of the case where information is transmitted from source (the company headquarters) to destination (the branch office).

While improvements in technology are facilitating the way information can be transferred, this brings along the issue of providing the security and privacy of information transferred. These issues are addressed in the security and privacy section that follows.

2.2 Information in process

This state focuses on how one operates on data to change its form. It is very common in computational operations today to process data such that it no longer possesses its original format. As an example, one may compress data so that it occupies less space, another may encrypt it so that it becomes unintelligible to unintended third parties, yet another combines compression and encryption so as to benefit from combined effects of the two.

Information in process is represented graphically in **Figure 2**, where a series of operations are applied to input data to yield the output data. The expectation in the end of the process in this figure is that the output ($Information_2$) will be different from input ($Information_1$). This is a fundamental difference from **Figure 1**, where information on each side was expected to be the same.

In the context of information in process, take, for example, the process of encrypting text. The input data will be a legible text, such as a sentence, a document, etc., whereas the output generated will be an unintelligible sequence of characters, as is deliberately meant to be by the process. The process itself could be one single operation as simple as circularly shifting characters a certain amount to

Figure 2.
Information in process.

the left or right or, in the hope to achieve sophisticated complexity, a sequence of operations, such as applying a complex math-based prediction model on previous data to calculate the overall end of year benefit for a company.

The concern with information in process is that the reliability of data should be preserved at all times so that the resulting data can be trusted. This involves authentication and reliability of input and output data, as well as the trustability of the process itself, each of which are discussed later in the Security Mechanisms for Three States of Information subsection below.

2.3 Information in storage

Information in storage state refers to the case when information rests in a storage media of choice for making it available in the future. **Figure 3** depicts the last state of information which is called information in storage.

The expectation at this state is to ensure that the information remains intact in the storage and moreover no unauthorized party can access to it. These efforts involve authentication of the users who can access the information, as well as integrity of information being stored. The requirement to keep the data intact, i.e., prevent its integrity while in storage, explains why the information is labeled the same in both user and storage sides in **Figure 3**. This is a similar requirement that was mentioned while explaining the information in transit, as was depicted in **Figure 1**. Yet, as information in process inherently changes the format of data, the labeling of information on each side of the process was different for information in process (see **Figure 2**).

As for the storage media, various options are possible: Information can be stored on a local database, or on cloud, particularly when local storage is not sufficient to accommodate large amount of data.

Once stored, information usually is accessed again for retrieval purposes, explaining the full duplex form of communication between the user and storage.

As an example, a person can store family vacation pictures on his cell phone and then may decide to store them on a DVD for allocating more space on his cell phone storage.

It should be noted that the distinction among these three states of information, namely, transit, process, and storage, is not crystal clear. This is because almost all the time these states are utilized in an overlapped manner. Formally, if the set of states is $S = \{s_1, s_2, s_3\}$, then all possible permutations of these states generate the following permutation set $P = \{\emptyset, s_1, s_2, s_3, s_1s_2, s_2s_1, s_1s_3, s_3s_1, s_2s_3, s_3s_2, s_1s_2s_3, s_1s_3s_2, s_2s_1s_3, s_2s_3s_1, s_3s_1s_2, s_3s_2s_1\}$. It should also be noted that it is permutations, not combinations, as the order matters when combining each of the three states. An example regarding the three states of information could be given as follows: Compressing data first and then emailing it to destination should be considered a different operation than emailing the data first and then compressing it at the destination.

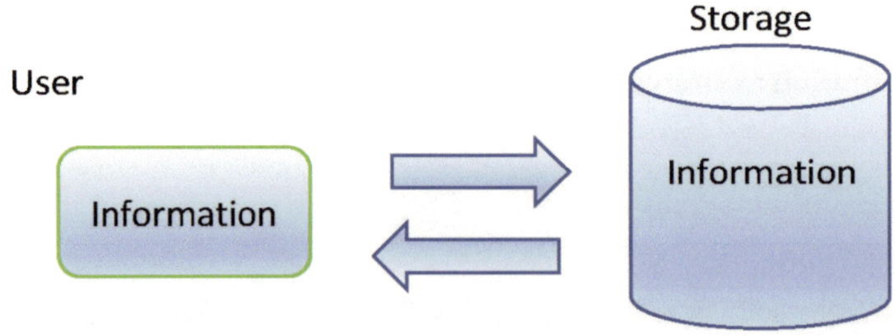

Figure 3.
Information in storage.

Permutation	State	Description
1	Ø	No operation (trivial case)
2	s_1	Transit
3	s_2	Process
4	s_3	Storage
5	s_1s_2	Transit-process
6	s_2s_1	Process-transit
7	s_1s_3	Transit-storage
8	s_3s_1	Storage-transit
9	s_2s_3	Process-storage
10	s_3s_2	Storage-process
11	$s_1s_2s_3$	Transit-process-storage
12	$s_1s_3s_2$	Transit-storage-process
13	$s_2s_1s_3$	Process-transit-storage
14	$s_2s_3s_1$	Process-storage-transit
15	$s_3s_1s_2$	Storage-transit-process
16	$s_3s_2s_1$	Storage-process-transit

Table 1.
Permutation of information states.

Table 1 presents all possible permutations of these three states applied either in isolation, or combined in pairs, or combined in triplets based on the underlying scenario each represents. The states listed in **Table 1** map to generic states s_1, s_2, and s_2 of the formal definition given above as follows: s_1 = information in transit, s_2 = information in process, and s_3 = information in storage.

Take, for example, the case where the husband retrieves previous year's tax records from his hard disk and emails them to his wife, and then the wife processes them to create this year's tax amount to be paid to the government. All three states of information are employed in this scenario in this particular order, storage-transit-process, corresponding to line 11 of **Table 1**. To make it even more complex, imagine the wife submits the tax form for this year via online submission and then stores it in her disk. This would mean expanding the order of states to include "transit-storage" operations in addition to what already exists in line 11 of the table.

3. Security mechanisms for three states of information

In this section, each of the three states of information described in Section 2 is addressed, regarding the security features each state requires. In particular, Section 3.1 focuses on the security of information in transit state and hence introduces and explains network security. Section 3.2 concentrates on the state of information in process and explains the security measures, algorithms, and their implementations defined under the title of computer security. And Section 3.3 analyzes the security of information in storage state and provides information pertaining to database as well as cloud security.

As each state handles information differently, the security requirements per state differ accordingly. For this reason, subsections below elaborate security concerns with regard to each state of information and present various techniques to provide the security of information in each of these states.

3.1 Network security

Different network infrastructures are probable today. The decision to make on which network infrastructure to employ depends on the needs for speed, reliability, the type of data being transferred, and the technology available where the source and destination are. The most common network infrastructures are TCP/IP and UDP communications. Each of these communication protocols utilize a layered architecture to process transmitted data. Essentially, they use similar fields for uniquely identifying the data being transmitted (such as source IP address, destination IP address, source port number, destination port number, flags, and other fields).

The fundamental security services that are expected to be provided by network security are confidentiality, integrity, availability (known as CIA principles), and authentication. In the following subsections, each of these services is discussed, and examples of tools and mechanisms used to provide these services are given.

3.1.1 Confidentiality

Confidentiality refers to the service where secrecy of the data being transmitted is preserved at all times. It is an essential requirement to ask from a network communication as otherwise the trust is violated irreparably.

The means that to provide confidentiality service is to employ encryption, i.e., encoding data so as to make it unintelligible to unintended third parties.

Based on the number of encryption keys used in an encryption system, two types of encryption are possible: symmetric encryption and asymmetric encryption. In a cryptosystem where C is the ciphertext, P is the plaintext, K is the secret key, E is the encryption algorithm, and D is the decryption algorithm, the symmetric encryption and decryption processes are formulized as seen in Eqs. (1) and (2), respectively.

$$\text{Symmetric encryption scheme encryption process: } C = E_K(P) \tag{1}$$

$$\text{Symmetric encryption scheme decryption process: } P = D_K(C) \tag{2}$$

In symmetric encryption, each communicating party uses the same secret key to encrypt the transmitted content. The need to use a secret key in symmetric encryption raises the question of how to provide the secrecy of this key itself. The solution is to employ a different key (named as session key) each time a new transmission takes place. Surely this adds on to the complexity of the scheme, but this is the inevitable result of the tradeoff between secrecy and complexity in security systems. The most common symmetric encryption algorithms are DES [1] and AES [2] with key sizes of 128 bits and 256 bits, respectively.

Take, as an example, the case where n users communicate with each other by using symmetric encryption scheme. To achieve a secure communication, each entity will have to ensure that his/her communication will not be revealed to other entities in the system. This will lead to a requirement to employ $\frac{nx(n-1)}{2}$ total number of symmetric keys in the system. Known as the n^2 problem [3], the size of symmetric keys becomes quadratically complex and is problematic particularly for large n. To alleviate the n^2 problem, a new type of encryption called asymmetric encryption has been introduced.

In asymmetric encryption, each communicating party possesses a key pair {privateKey, publicKey} and employs this pair for communicating with any of the other parties in the system. Therefore, for communication among n users, the need for total number of keys is 2n only, a dramatic reduction as compared to the symmetric key encryption. A common asymmetric algorithm is RSA that is widely used in cryptographic platforms.

Another classification for cryptographic algorithms is based on how the plaintext bits are processed: stream cipher and block cipher. In stream cipher, as the name implies, the plaintext bits are processed individually to encrypt, while block cipher bits are processed in blocks of multiples of 8 during encryption. The inherent nature of these encryption methods introduces a tradeoff between the two: Stream cipher is faster and has the advantage of low error propagation, yet it is more prone to malicious bit injection as they can go undetected very easily. The block cipher performs slower, as it has to incur the extra cost of block forming each time. It also cannot prevent error propagation in blocks. Yet, the block cipher is secure against malicious bit injection.

Yet another classification for cryptographic algorithms is based on how the plaintext is transformed into ciphertext. According to this classification, one can perform a substitution cipher or transposition (permutation) cipher. In substitution cipher, plaintext bits are transformed into ciphertext bits by employing a single alphabet (hence called monoalphabetic cipher) or multiple alphabets (hence called polyalphabetic cipher). In a monoalphabetic cipher, the key size is fixed and is reused for the entire plaintext encryption, which leads to a 1:1 mapping between plaintext and ciphertext letters. As an example, for encrypting a given plaintext in English with length n, if the encryption key is designated as K = 5, then the encryption algorithm could be written as seen in **Figure 4**.

Similarly, the corresponding decryption algorithm for monoalphabetic cipher can be written as seen in **Figure 5**.

Monoalphabetic cipher is straightforward to implement. The risk is once the key is known, it will lead to decryption of the entire ciphertext promptly, violating the requirement for security. As for deciphering a suspected to have been monoalphabetically encrypted text, one can use the source language (such as English) n-gram statistics to reveal the key size. The oldest known encryption algorithm Caesar cipher is a monoalphabetic cipher with a key size of 3. Once statistical analysis is used to decrypt monoalphabetic cipher, the letter frequencies of ciphertext will reflect the letter frequencies of source language, only with "substituted" letters this time. As an example, if key size is 4 for a monoalphabetic cipher, then the

```
monoalphabeticEncrypt(P[], K)
for i=1 to n do {
        C[i] = P[i] + K;
        print (C[i]);
}
```

Figure 4.
Monoalphabetic cipher encryption.

```
monoalphabeticDecrypt(C[], K)
for i=1 to n do {
 P[i] = C[i] - K;
        print (P[i]);
}
```

Figure 5.
Monoalphabetic cipher decryption.

plaintext letter “A” will be encrypted into ciphertext letter “E,” ciphertext letter “B” will be encrypted into plaintext letter “F,” etc. Therefore, the frequency of the ciphertext letter “E” will be similar to the frequency of source language letter “A,” the frequency of the ciphertext letter “F” will be similar to the frequency of source language letter “B,” etc. This is a very important clue in helping one decrypt the ciphertext correctly and fast.

In an effort to strengthen the security of monoalphabetic cipher, polyalphabetic encryption has been introduced. As the name implies, in polyalphabetic cipher, there is an M/N mapping between plaintext letters and the ciphertext letters. This means that a plaintext letter “A” may be encoded into cipher letter “Y” once, and into ciphertext letter “F” another time, and into ciphertext letter “K” yet another time, etc. Therefore, there is no one fixed key size which will make decryption harder and more time-consuming. This will also cause a phenomenon called uniform letter frequencies for the ciphertext letters, which will be very close to uniform distribution of letters, rather than reflecting the natural language statistical characteristics (where letters “A,” “E,” and “T” occur the most frequent, whereas letters “Z” and “Q” occur the least frequent for English). Deciphering a ciphertext with uniform letter frequencies is very hard as compared to deciphering a ciphertext that possesses the underlying source language frequencies.

In the case of a polyalphabetic cipher, more complex deciphering techniques such as Kasiski test [4] and index of coincidence [5] should be used. Both these techniques employ statistical analysis on ciphertext partitions formed by reordering ciphertext letters in varying lengths to reveal a meaningful plaintext correspondence, in the hope to find a most probable key size. The ultimate polyalphabetic encryption algorithm is called the Vernam Cipher (also called one-time pad) [6] and uses a key size as large as plaintext to obstruct key size computation and prevent the facilitating contribution of n-gram statistics.

The transposition (permutation) cipher uses the same letters as the plaintext while encrypting the text. As an example, in rail cipher where key size is K, the plaintext letters are combined by right shifting each letter K positions to form the ciphertext. Therefore, the ciphertext letter frequencies will remain exactly the same as the plaintext letter frequencies. This is a very important discovery that will help determine whether a transposition cipher or a substitution cipher was used in the first place.

Regardless of the type of encryption algorithm, confidentiality service for network security is provided by applying the encryption cipher of choice. For a system with symmetric encryption, encryption algorithm for the sender side will be as seen in Eq. (3), and decryption algorithm for the receiver side will be as seen in Eq. (4):

$$\text{Symmetric encryption sender side } C = E_{K_1}(P) \tag{3}$$

$$\text{Symmetric decryption receiver side } P = D_{K_1}(C) \tag{4}$$

For a system with asymmetric encryption, each entity will use a pair of keys {privateKey, publicKey}. Hence, encryption algorithm for the sender side will be as seen in Eq. (5), and decryption algorithm for the receiver side will be as seen in Eq. (6):

$$\text{Asymmetric encryption sender side } C = E_{ReceiverPublicKey}(P) \tag{5}$$

$$\text{Asymmetric decryption receiver side } P = D_{ReceiverPrivateKey}(C) \tag{6}$$

More security services are needed to provide a comprehensive security for a communication network. In the following section, these additional services are introduced.

3.1.2 Integrity

While confidentiality focuses on the secrecy of data, integrity overlooks this concern and is primarily concerned about the trustworthiness data, i.e., the data is not changed by unauthorized entities. Consider a banking transaction where 1000$ is transferred from one account to another. The identity of sender and customer may be known to those who are processing this transaction. So, confidentiality is not a concern. Yet, the fact that the 1000$ amount should be transferred fully and correctly is the upmost concern and should be addressed via integrity.

In a network setting, integrity is achieved by implementing a hash algorithm on data to be transmitted. A hash algorithm H is an encoding function that takes an input (plaintext) P of any size to yield an output D (digest) of fixed size as seen in Eq. (7):

$$D = H(P) \tag{7}$$

Implementing a hash algorithm on input data generates a message digest, which is then transmitted to the receiver along with the plaintext. It should be noted that for a cryptographic system where integrity is the only concern, confidentiality is disregarded and plaintext can be sent in the clear. Having received the plaintext, the receiver applies the same hash function (that should have been agreed upon beforehand) on the plaintext and compares the computed digest with what has been received. If they match, it means that the integrity of data has remained intact. Otherwise the data has been compromised in transit.

In order for hash function to perform properly, it should have all two of the following properties:

- One-way property: given a digest $D = H(P)$, calculating plaintext $P = H^{-1}(D)$ should be computationally difficult. To accomplish this goal, one-way functions exploiting computationally hard problems (such as discrete logarithm problem) are designated as hash functions.

- Weak collision property: for two different plaintexts as P_1 and P_2, where $D_1 = H(P_1)$ and $D_2 = H(P_2)$, the probability of $D_1 = D_2$ should be very small. It should be noted that regardless of plaintext size, the digest has a fixed length. So, a collision is more likely to occur than is actually anticipated. Once this is the case, techniques such as double hashing, linked list, etc. are applied to avoid collision with an overhead of extra process and/or space.

Several hash functions having strong hash properties exist to provide integrity of data. Examples of such functions are SHA [7], MD5 [8], etc.

3.1.3 Availability

For confidentiality to provide secrecy, and for integrity to provide trustworthiness of data, availability of this data should be guaranteed first. In network security context, availability is concerned about the entities being available at all times throughout communication. Those entities are information pertaining to the sender and receiver, the data itself that is being sent, and the metadata that is uniquely identifying the data under operation.

Attacks targeting fields of network protocol headers (IP, TCP, MAC headers) may disrupt availability. Denial of service (DoS) is a common type of network security attack that exploits the three-way handshake protocol between a sender and a receiver. Ideally, the three-way handshake should start by step 1 where the sender sends the TCP header with SYN flag field set along with a sender sequence number. Upon receipt of this request, in step 2 the receiver—if available—informs its availability by sending back the TCP header with ACK flag field set, the sender sequence number + 1, and the receiver sequence number to the sender. Finally, in step 3, the sender sends the TCP header with ACK field set, sender sequence number + 1, and the receiver sequence number + 1. Only then the actual communication can start.

When the three-way handshake is compromised, a rogue sender bombards a victim receiver by sending him too many "half-open" connections, i.e., step 1 of the three-way handshake protocol with SYN flag fields set in each with a different sender sequence number. The victim receiver tries to respond to each SYN request by sending back an ACK + SYN response addressing each independent sender sequence number, but as the rogue sender deliberately never sends back an ACK response to these responses, the victim gets overwhelmed shortly as its half-open connection buffer overflows. The solution to DoS attack is to limit and advertise the buffer size for each entity in the network.

Other examples of network attacks targeting availability can be listed as SYN guessing, IP spoofing (by impersonating a legitimate entity), covert channels (by embedding secret data on unused fields of network protocol headers), messing up fragmentation of packages so that they will not defragment correctly in the receiving entity, etc.

A common protection against availability attacks is to use firewalls. They protect systems from outside entities by enforcing strict rules only to allow packets with particular properties (IP number, port number field, etc.).

3.1.4 Authentication

Authentication focuses on whether the originator of data really is who he claims to be. It is directly associated with trustworthiness, which makes it closely related to integrity. In other words, authentication is origin integrity.

To provide authentication in network systems, digital signature is used. Digital signature is an encoding function that is similar to a hash function, with the additional property of having a key. The key should belong to the originator, as an evidence for proving his identity. A hash function with a key is called message authentication code (MAC). Several MAC algorithms exist, HMAC [9] being the most popular.

In a digital signature scheme, asymmetric encryption is used, where encoding and decoding are named as signing and verifying, respectively. Eqs. (8) and (9) list the signing and verifying algorithms for digital signature.

$$\text{Signing algorithm: } C = S_{SenderPrivateKey}(P) \tag{8}$$

$$\text{Verification algorithm: } P = V_{SenderPublicKey}(C) \tag{9}$$

ElGamal algorithm is a commonly used digital signature scheme. It provides varying levels of authentication based on the choice of key size.

3.2 Computer security

The concentration on computer security is on secrecy and privacy of data during processing. Therefore, those security tools and mechanisms introduced for network security also apply to computer security. In particular the following three security services are needed to deliver a comprehensive security service while processing data:

- Cryptography to help with confidentiality
- Hash functions to provide integrity
- Digital signatures to help with providing authentication

Take, as an example, a banking transaction in which 1.5% interest is added to customer's bank account. It is very important that identity of this account holder should be kept confidential. So, an encryption tool should be incorporated to provide confidentiality. Moreover, the balance amount should remain intact throughout transaction (except when the resulting balance is calculated), and this requirement can be met via hash algorithms. Furthermore, the system should assure at all times that the transaction belongs to the actual owner of the account, but no other account holder. So, a digital signature scheme should also be employed to prevent repudiation, hence to provide authentication.

3.3 Database/cloud security

As being one of the probable states information can reside in, storing data usually involves one of two media: database or cloud.

3.3.1 Database security

When data is stored on a database, the security measures that should be considered comprise of the following:

- Encrypting the data stored so as to provide confidentiality.
- Compressing data so as to occupy less space—additionally, compression offers augmented security as it shuffles data. A large selection of compression algorithms is available. The choice on which algorithm to choose depends on data type (for text data, a lossless compression algorithm should be utilized, while for image files, voice and video lossy compression can be tolerated), space requirements, speed, and compression rate (how much the compression algorithm reduces the input size).
- Prevent against SQL injection attack so that input forms for data entry cannot be exploited to potentially damaging SQL instructions. As an example, if a database system is vulnerable against SQL injection attack, an attacker can "inject" a rogue statement (such as dropping a database table) to execute, immediately following a biased statement (e.g., as simple as "1 = 1") that will always yield true. A suggested method to prevent against SQL injection attack is to use data sanitization (so that data entry will not allow some characters such as apostrophe, etc.), or using stored procedures to execute instructions, rather than exposing forms for easy injection. There exist several forms of SQL injection, and cross-site scripting (XSS) is one of them.
- Secure the database physically so that data can be protected against access from unintended third parties.

3.3.2 Cloud security

Based on NIST's cloud security definition [10], cloud provides one of the three fundamental services:

- Infrastructure as a service (IaaS) allows subscribers to execute any application and OS on the hardware and resources (abstracted via hypervisors) made available by the cloud. Some examples of IaaS type are Amazon Web Services (AWS), Google Compute Engine (GCE), Microsoft Azure, Rackspace, and Cisco Metapod.

- Platform as a service (PaaS) allows subscribers to create their custom applications on the cloud. The cloud makes itself available to its customers by providing tools such as a DBMS, OS, system software, and applications. The examples of PaaS type can be listed as Apache Stratos, Windows Azure, OpenShift, Heroku, Google App Engine, and AWS Elastic Beanstalk.

- Software as a service (SaaS) subscribers sign a service agreement for this service to execute cloud-owned online applications. Some common examples of SaaS type of cloud service are listed as follows: GoToMeeting, Salesforce, Dropbox, Cisco WebEx, Google Apps, and Concur.

Regardless of the type of service cloud provides, security of the cloud should consider boundary security that builds itself on the layered architecture of hypervisors. Moreover, hardware boundary and abstraction boundary definitions are protections should be offered regarding whether the cloud itself is private or public.

It is dramatically important for cloud to isolate individual customers' data so that they will not interfere with other customers' data. So, anonymity is a primary concern. A straightforward technique to provide anonymity is by incorporating encryption.

Moreover, it should be computationally infeasible to extract summary data that will bring together multiple subscribers' data pool, as this may lead to unfair and unethical advantage. As an example, personal healthcare data stored for multiple healthcare providers should not be analyzed easily to extract a conclusion that persons inhabiting in a particular region are more prone to a particular disease as this may cause people from this region be charged higher by insurance companies.

Though security measures are classified and analyzed separately, due to the complex nature of information systems, handling information most of the time involves multiple aspects of security at the same time. For this reason, complex information security systems have been developed and are widely used. Some examples of such systems are Pretty Good Privacy (PGP) [11] for data encryption, integrity, and authentication, Kerberos for secure key distribution, and many more [12].

4. Conclusions

This section introduces the three fundamental states of information, namely, information in transit, information in process, and information in storage, and then discusses the security measures pertaining to each state of information. In particular, network security methods to provide security of information in transit, computer security methods to provide security of information in process, and database/cloud security to provide security of information in storage are introduced and discussed in detail.

As the computer technology evolves in rapid speed, security measures will be extended by either improving the existing algorithms or adding more algorithms to the suit.

Author details

Ebru Celikel Cankaya
Department of Computer Science, University of Texas at Dallas, Richardson, TX, USA

*Address all correspondence to: exc067000@utdallas.edu

References

[1] Coppersmith D. The data encryption standard (DES) and its strength against attacks. IBM Journal of Research and Development. May 1994;**38**(3):243-250

[2] Daemen J, Risjemn V. The Design of Rijndael. New York, USA: Springer-Verlag; 2002. p. 1

[3] Bishop M. Computer Security: Art and Science. Boston, MA, USA: Addison-Wesley; 2003

[4] Pommerening K. Kasiski's test: Couldn't the repetitions be by accident? Journal Cryptologia. 2006;**30**(4):346-352

[5] Friedman WF. The Index of Coincidence and its Applications in Cryptanalysis. Walnut Creek, CA, USA: Aegean Park Press; 1987

[6] Vernam G. Cipher printing telegraph systems: For secret wire and radio telegraphic communications. IEEE. February 1926;**45**(2):109-115

[7] NIST SHA-3 Standard. Permutation-Based Hash and Extendable-Output Functions. 2015. DOI: 10.6028/NIST.FIPS.202

[8] Guzman LB, Sison AM, Medina RP. MD5 secured cryptographic hash value. In: International Conference on Machine Learning and Machine Intelligence, September 2018. pp. 54-59

[9] NIST. The Keyed-Hash Message Authentication Code (HMAC). Available from: https://csrc.nist.gov/publications/detail/fips/198/1/final

[10] NIST SP 800-145. The NIST Definition of Cloud Computing. 2011. Available from: http://csrc.nist.gov/publications/PubsSPs.html#800-145

[11] Choi YB, Hunter ND. Design, release, update, repeat: The basic process of a security protocol's evolution. International Journal of Advanced Computer Science and Applications. 2017;**8**(1):355-360

[12] Stallings W. Cryptography and Network Security. 7th ed. New York, USA: Pearson; 2016

Chapter 5

The Role of Gamification in Privacy Protection and User Engagement

Aikaterini-Georgia Mavroeidi, Angeliki Kitsiou and Christos Kalloniatis

Abstract

The interaction between users and several technologies has rapidly increased. In people's daily habits, the use of several applications for different reasons has been introduced. The provision of attractive services is an important aspect that it should be considered during their design. The implementation of gamification supports this, while game elements create a more entertaining and appealing environment. At the same time, due to the collection and record of users' information within them, security and privacy are needed to be considered as well, in order for these technologies to ensure a minimum level of security and protection of users' information. Users, on the other hand, should be aware of their security and privacy, so as to recognize how they can be protected, while using gamified services. In this work, the relation between privacy and gamified applications, regarding both the software developers and the users, is discussed, leading to the necessity not only of designing privacy-friendly systems but also of educating users through gamification on privacy issues.

Keywords: privacy, security, privacy requirements, privacy awareness, game elements, gamification

1. Introduction

Due to the digitalization of information, the use of several technologies has been introduced in people's habits, which, consequently, signifies the prevalence of applications utilization, pertaining to many sectors [1]. The Information and Communication Technologies (ICTs) may have different scopes and concepts based on the preferences and the aim of their developers. A variety of such technologies [2, 3] have been provided, aiming at educating users on specific topics, for example, by educational platforms for students and teachers, at endorsing products for marketing purposes, or other reasons, depending on the concept of each application.

Specific techniques or methods have been developed in order to improve the provision of their concept, such as a more appealing and interactive environment [4]. By achieving to gain this benefit, a service will be more interesting, and its use will be increased. Consequently, the aim of its developer will be satisfied, as for instance in marketing domain, this will support the company's profits. An example of such methods is gamification [4], the use of game elements in applications that are not games. Many game elements have been recorded in the literature and have been implemented in gamified services [5–11] in order to support several domains [7, 8, 12–16]. The benefits, provided by the implementation of the game elements, differ based on the concept of each service. The singularity of this method concerns on the increased engagement of users with gamified environments [4].

According to the above, it is clear that these technologies consist of a basic activity for users, and some of them may be helpful for their lives, for example, educational platforms. However, there are some issues that arise, namely security and privacy issues. While using all these ICTs, users' information is stored, and their activities are recorded [17–19]. As a consequence, users' personal information may be harmed. Thus, it is crucial to consider the protection of users' security and privacy while designing services. Regarding the relation between gamification and these two important aspects, some studies have been published [13, 14, 20, 21], in which the harmful side of gamification, focusing on users' security and privacy, is presented. Besides the importance of designing privacy-friendly and secure gamified services, it is also crucial for a user to be able to protect his/her own security and privacy. If a user becomes aware of this, then, several ethical and social issues will be addressed. So, on the one hand, it is important for developers to design security and privacy-friendly gamified services and on the other hand, users should be security- and privacy-aware. Adding to this point, in this work, the importance of users' privacy awareness on protecting their privacy through gamification is discussed, and some preliminary results are presented.

The rest of the chapter is organized as follows. In Section 2, gamification is described, providing its benefits and implementation on several sectors. In Section 3, the relation between gamification and privacy is presented. Additionally, the importance of security and privacy awareness regarding gamification is highlighted and the contribution of gamification on educating users on privacy is discussed. Finally, Section 4 concludes the work, providing steps for future work.

2. Gamification in ICTs

Around 2010, the method of gamification has been introduced in ICTs, aiming to engage users on using technologies and to increase their interest [22]. By implementing this method and especially by introducing game elements, the main principle of gamification, namely a more gameful interaction environment, can be developed [23]. The definition, which is highly cited in previous research, was published in 2011 by [4], who defined gamification as the use of game elements in nongame contexts. Many game elements have been presented in the literature and their choice depends on the developers' scope, the concept, and the structure of the gamified service. Each study mentions and describes an amount of game elements [24–26]. In [27], all mentioned game elements have been presented by conducting a review that recorded all game elements and introduced them in the relevant literature. In **Table 1**, the amount of game elements is presented along with the explanation of their concept. Additionally, some examples of gamified applications in several sectors are given out with the respective elements that have been assigned to them. In this work [27], the connection among elements can be identified. For instance, in order for a user to win badges, levels have to be passed or points have to be collected [39]. So, with the intention of an application to be gamified, it is usual to include many game elements.

The gamification method has been implemented in several sectors. Starting from the sector of education, many gamified services provide a more entertaining environment, which automatically gains users' engagement. Therefore, users can be educated on different topics, without having in mind the literal sense of the education process.

Game elements	Explanation	Examples of studies or gamified services
Alternative activities	Many provided choices and tasks to users	[28, 29]
Achievements	The accomplishment of a task	[29–31]
Avatars	Users' representation through animated processes	[31, 32]
Badges	After winning or accomplishing a task, badges are given	[12, 22, 28–38]
Challenges	The ability of a user to challenge a friend in order to compete	[5, 12, 22, 28, 31, 33, 39–41]
Communication with other players	Users' communication through respective platforms	[12, 28, 30, 32]
Competition	Users' competition on some steps	[12, 29, 31, 32, 34, 39, 42]
Content unlocking	Steps that have to be passed in order to unlock the next phase	[31]
Feedback and progressive information	Provided information to help users for their status and recommendations	[22, 28, 30–33, 39, 42]
Leaderboards	Users' status on the service regarding their points or level	[12, 22, 28, 29, 31, 33–35, 39, 42, 43]
Levels	Phases that have to be passed	[12, 22, 28–32, 35, 39, 42]
Location	The connection with users' location	[28, 32]
Notification	Users are notified to accomplish actions	[28, 32, 39, 40]
Points	The result of finishing a task can be illustrated by the collection of points	[5, 22, 28–37, 39–42]
Profiles	Each user has his own profile on the service	[28, 32, 39, 40, 42]
Quiz	Questions on a specific topic	[29, 32, 40, 42]
Rewards	The result for winning an opponent or effectively completing a task	[5, 12, 22, 29, 39–42]
Roles	The character that a user wants to have	[28, 29]
Rules	The dos and don'ts that users have to follow	[28, 29, 31, 32, 39]
Scoring systems	Systems which record users' score and status	[12, 28]
Team tournaments, group tasks and collaboration	Tasks where users have to collaborate	[28, 29, 31, 32, 39, 40, 42]
Time constraints	Actions that have to be completed during a specific time period	[12, 28, 29]

Table 1.
The recorded game elements.

In some educational services, users can be either teachers or students, where teachers provide feedback and communicate with students. Communication can be also achieved among students, while having the opportunity to compete with each other [42]. In this way, apart from the knowledge benefits, it is also important that users' sociability can be expanded. In marketing domain [10, 12, 44], the aim of this method is to raise each company's selling. By providing applications, where users collect points after buying a product with the deal to win a gift card or a product, the application can be further used. Gamification in this domain is a smart idea so as to engage users and sell more products. Additionally, in some cases, the interaction among users is enhanced, either through competition or collaboration, leading to users' amiability.

The role of gamification in health domain is quite crucial [23, 45–47]. The aim of the most gamified healthcare services is to educate users and engage users on protecting their health. A variety of such services can be found either for children or for adults. Some of them provide the opportunity of interaction between doctors and consultants, where, as an example, doctors can monitor the patients' progress on taking their medication [40]. Most of them notify users every time they have to take their prescription [40]. The gamified principle in such services can be the collection of points after responding to doctors' advices and prescriptions, resulting, sometimes, in the win of gifts. Therefore, users can protect their health through a more entertaining process. Beyond the above sectors, gamified services have been developed for cultural [11, 25] or touristic purposes [10, 48] offering benefits, such as cultural education and tourism's expansion, respectively. Furthermore, some studies tried to combine gamification with software engineering, indicating the state of the art on this field and the research gaps [49], while others elaborated research on gamification and education on software engineering in order to identify the discussed works [8].

To a lesser extent, studies which concern on gamification and security have been recorded [13, 14], aiming to highlight the important role of security in services. Apart from the importance of gamification in security, it is also crucial to educate users on privacy issues, since by using these services, users' information is often disclosed. However, few research attempts have been identified, which combine gamification and privacy [19]. A more detailed analysis regarding gamification and privacy has been provided by [17], who focused on the software aspect of gamified applications regarding users' privacy. They identified that gamification is a method, whose principles may harm privacy requirements. Especially, in [26], a metamodel has been published, aiming to point out how privacy violation can be achieved by the core of gamification, in particular, the game elements. However, studies regarding the importance of users' awareness on privacy issues, as in the security area, have not been recorded yet, which is a crucial research gap.

According to the literature [22, 26], gamifying a service is a useful process for many reasons, discussed above. Since it consists of a method that has been introduced in ICTs the last years, more research is needed to be conducted concerning its relationship with other sectors, such as privacy and security.

3. Examining privacy in gamified services

Although privacy is an aspect that should be considered during the design phase of each type of service, it has been identified that few researchers have been focused on its relation with gamification. According to the literature, privacy satisfaction is based on the analyzation and elicitation of privacy requirements on the systems [50]. Many relevant engineering methodologies have been published which describe these requirements and explain how they can be analyzed within the systems [13, 17, 18, 50–58]. In [58], all requirements that were mentioned and used in [27] for the conduction of the results are described. These requirements are presented in **Table 2** along with their aim. An in-depth combination between gamification and this aspect, focusing on the peculiarity of gamified services related to the privacy requirements is provided in [27]. This relation has been examined, paying particular attention to the impact of gamification on privacy domain. Specifically, they recorded all game elements reported in the literature and identified which of them may harm users' privacy. This identification was based on the concept and the scope of each game element. Based on their findings, specific elements are identified whose concept is harmful for privacy requirements. In **Figure 1**, the

Privacy requirements	Aim
Anonymity	The identity cannot be compromised
Pseudonymity	The use of a pseudonymous to ensure identity's anonymity
Unlinkability	The actions and identities cannot be linked
Udetectability	The existence of a component cannot be detected
Unobservability	The actions The actions between identities cannot be observed

Table 2.
Privacy requirements.

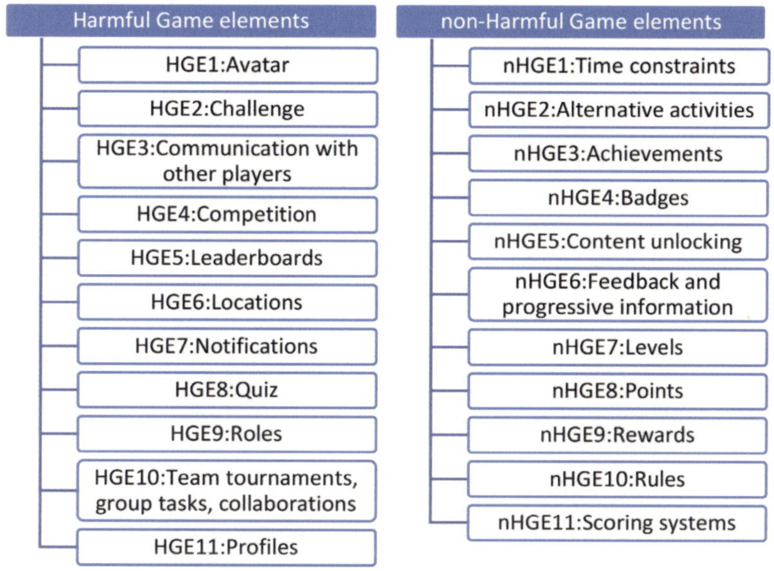

Figure 1.
Harmful and nonharmful game elements for privacy. HGE, harmful game element; nHGE, non-harmful game element.

categorization among harmful and nonharmful game elements is presented. For instance, when using a service which records users' personal information (HGE11), location (HGE6), and his/her interaction with other users (HGE2, HGE3, HGE4, and HGE10), then user's privacy cannot be protected. On the other hand, the selection of points (nHGE8) in order to pass levels (nHGE7) or the rules (nHGE10) and time constraints (nHGE1) are not harmful game elements, as, for instance, user's information or actions are not recorded due to the constraint of time.

Afterwards, a more detailed analysis has been published in [26], where authors presented their findings by designing a metamodel. In detail, after the first investigation of the relation between game elements and privacy requirements [27], authors selected some existent gamified services and recorded the used elements in order to examine their findings on real environments. According to the results [27], the game elements that have been implemented in these gamified services may harm users' privacy by violating the privacy requirements. The findings are illustrated in a metamodel which presents how each element is in conflict with the privacy requirements [26]. This conflict arises from the identified disadvantages of the game elements in [27]. Expanding previous work and according to this way of examination, in **Table 3**, the relation between game elements and privacy

Game elements	Reason of violation	Violated privacy requirements
Avatar	Recognition and recording of user's characteristics	R1, R2, R3, R4, R5
Challenge	Recognition of the opponent's information and connection between identities	R1, R3
Communication with other players	Recognition of the user's characteristics and interaction between identities	R1, R2, R3, R4, R5
Competition	Recognition of personal information and connection between identities	R1, R2, R3, R4, R5
Leaderboards	Recognition and recording of the opponent's information	R1, R2, R3
Location	Recording of user's location	R1, R2, R3, R4, R5
Notification	Recording user's actions depending on his reaction	R1, R2, R3, R4, R5
Quiz	Recording of user's awareness and information	R1, R2, R3
Roles	Recognition of the user's preferences and behavioral characteristics	R1, R2, R3, R4, R5
Team tournaments, group tasks, collaboration	Recording and recognition of the user's interaction and information	R1, R2, R3, R4, R5
Profiles	Recording of user's personal information and connection with their actions and preferences	R1, R2, R3, R4, R5

R1, anonymity; R2, pseudonymity; R3, unlinkability; R4, undetectability; R5, unobservability.

Table 3.
The relation of game elements and privacy requirements.

requirements is presented. In particular, the game elements, presented in the metamodel have some advantages, which it is noted that at the same time are turned into disadvantages and consist the reason of their conflict with requirements. The disadvantages of the elements concern on the violation of (a) users' anonymity and (b) pseudonymity, due to the record of personal characteristics, preferences, and information, (c) the unlinkability and (d) undetectability of actions and identities, as actions are recorded and monitored in parallel to the identities, and (d) the unobservability, since by recognizing the identity and the actions, a third party can monitor them. For instance, even if "avatars" is an element which provides an animated representation of the user, the technique which is implemented to achieve that is the one of the face recognitions. In case, users' faces, that is, users' characteristics, are recorded, their identity can be compromised, so their anonymity can be violated, as the actions can be linked to this identity.

Based on the results published in [26, 27], gamification is a method which should be considered in parallel with privacy issues during the design of systems, since several game elements are harmful for privacy requirements. Despite the adequate number of published privacy engineering methodologies, it would be useful to combine the concept of them with the principles of gamification, so that privacy is protected in gamified services. Thus, a more comprehensive analysis regarding the recommended steps of these methodologies in relation to the game elements would be useful, in order to identify if and how they can be implemented on gamification processes. In addition, focusing on the privacy aspect, in [59], privacy patterns have been published which present how privacy requirements can be protected when developing a system. Such software patterns are important to be developed in relation to game elements in order for the software developers to implement them during the design of gamified systems. The design of privacy-friendly gamified systems is crucial, likewise the education of users on privacy issues. By this way, users will be able to use systems which protect their privacy, while in parallel, users' will be aware of how they can protect their privacy on their own.

Although it is important to provide gamified services which respect users' security and privacy, the crucial role of privacy and security awareness is undisputed. Especially, under the GDPR regulation, it is very important for users to know in which processes and why they give their permissions while using all ICTs. Information control is recognized as a key element in the perception and assumption of privacy risks [60]. Since, during the last years, many users use several types of ICTs to support their habits, the need of their awareness in order to protect their safety and personal information is increased even more. In [61], authors have published processes for the development of security awareness and training programs (SAT programs). The aim of these programs is the comprehension of security rules and the acquisition of skills regarding security, so as users avoid security violations that harm both themselves and the systems. For the development of SAT programs, four phases are recommended in [61]. The "Design phase" is the first step, where the budget, the target group, the needs of this group, and the program schedule have to be identified. The "Development phase" includes the determination of the concept and the issues that users should be aware of, for example, the protection of users' passwords and threats related to users' vulnerabilities. The third phase is the "Implementation phase", where the SAT program has to be implemented. In this step, it is important to explain the program to users in order for its purpose to be understandable. The "Post-Implementation phase" aims to record the use of the program for possible needed improvements, vulnerabilities, and advantages of it. Through a system, the results of its use should be recorded, so as the administrators of the program are able to monitor it during its implementation. Questionnaires, interviews, and other methods of evaluation are recommended for future improvement of the program.

Likewise, for the security issues, it is also important for users to be aware on privacy issues, so as to protect their personal information and actions. In order for a user to achieve his/her own protection, he/she has to be aware of some issues, such as if other users know their information, by whom, how, why, and which of the information can be distributed [62]. Users' privacy protection is ethically, legally, technically, and socially very important, for the sake of addressing any social harmfulness, deriving from privacy violation. For instance, cyberbullying, related to the disclosure of personal information, is a social phenomenon observed mostly in young people and concerns on users' harassment and unauthorized use of their personal information [63]. In accordance to this example, several respective phenomena arise by violating privacy, and therefore, privacy awareness is a crucial aspect in order to address them.

According to the findings of [26, 27], described above, there are game elements which harm privacy requirements. Thus, between privacy and gamification, the conflict concerns only the harmful game elements, as presented in **Figure 2**. By designing educational gamified systems, which provably [7, 12] engage users, with their concept to be on privacy issues, users will be able to protect their selves. In this figure, the major entities of privacy and gamification are illustrated, where on privacy domain the analyzation of privacy requirements in systems is needed to protect users' privacy, while in gamification, the design of gameful environments is crucial for the engagement of users. The relationship among entities is indicated and represented by directional bows that lead on the educational role of gamification in order for users to be aware of privacy issues. By adopting the harmful relation of these two entities [26, 27], which concerns on the harmful elements for privacy requirements, users can be trained on this, so that they will be educated (a) on the importance of

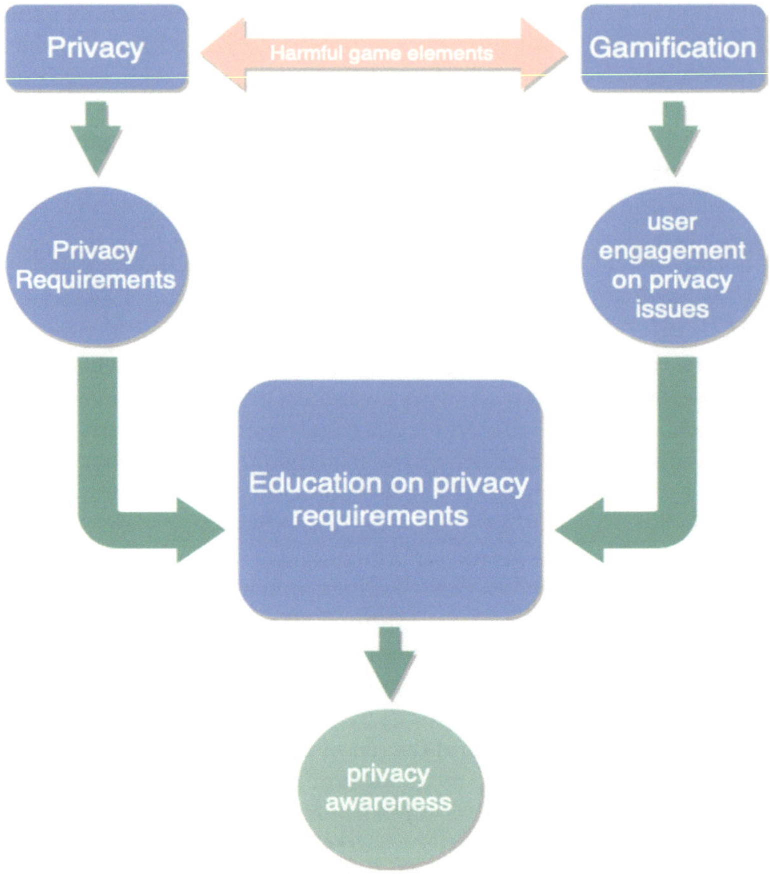

Figure 2.
Privacy protection by harmful game elements.

protecting the privacy requirements, (b) on recognizing the harmful game elements, (c) on how to protect their privacy while using these elements, and (d) on the consequences of their privacy violation if these elements carry on harming privacy requirements. The result of this process concerns the existence of awareness of users on privacy issues. Such educational programs, aiming to enhance users' privacy awareness level, are therefore significant in achieving a balance between users' need for the protection of their personal information during using gamified services and their need for using game elements within them that are harmful for their privacy.

Thus, in order to spread awareness to users through more entertaining processes, gamification can be considered, while developing privacy awareness services as well. Some examples have been recorded regarding security awareness [64, 65], but gamified attempts are also needed as far as privacy awareness concerns. The contribution of gamification in these services concerns on the engagement of users on using them, resulting on the effective education of users.

4. Conclusion

The implementation of game elements in ICTs is undeniably an effective way to engage users on using them. Several domains utilize gamification for the achievement of their purposes and many users prefer them for their tasks. While using them, personal information and actions are recorded, and therefore, privacy is an aspect that should be considered during developing them. Several respective methods which analyze security and privacy requirements have been recorded, but few attempts which combines them with gamification have been published. In this work, the relation between privacy and gamification is discussed, where it was highlighted that privacy may be violated by gamification. However, it is equally important for users to have awareness on privacy and security, so as to be able to protect themselves. Related programs which educate users on these two aspects are needed. Their combination with gamification is important in order for the users to be trained through a more interesting way. Some attempts have been recorded. In this work, it was identified that privacy awareness can be achieved by designing gamified systems which educate users on how to protect their privacy by the harmful game elements and on the consequences of privacy requirement violation. Thus, while using gamified systems, users will be able to know as on which game elements should pay attention in order to protect their privacy as how to be protected by their harmful consequences. In future work, software patterns for designing security- and privacy-friendly software will be recommended. As far as of the users' concern, privacy and security awareness will be studied in relation to gamification. In our purposes, the relation between gamification, security, and privacy is important to be examined as from the side of software developers as from the side of users.

Acknowledgements

The research work was supported by the Hellenic Foundation for Research and Innovation (HFRI) under the HFRI PhD Fellowship grant (Fellowship Number: 671).

H.F.R.I.
Hellenic Foundation for Research & Innovation

Conflict of interest

The authors declare no conflict of interest.

Author details

Aikaterini-Georgia Mavroeidi, Angeliki Kitsiou and Christos Kalloniatis*
Privacy Engineering and Social Informatics Laboratory, Department of Cultural Technology and Communication, University of the Aegean, Mytilene, Lesvos Island, Greece

*Address all correspondence to: chkallon@aegean.gr

References

[1] Conole G, Dyke M. What are the affordances of information and communication technologies? Research in Learning Technology. 2004;**12**(2):113-124

[2] Seth A, Vance JM, Oliver JH. Virtual reality for assembly methods prototyping: A review. Virtual Reality. 2011;**15**(1):5-20

[3] Azum AR. A survey of augmented reality. Presence. 1997;**6**(4):355-385

[4] Deterding S, Dixon D, Khaled R, Nacke L. From game design elements to gamefulness: Defining 'gamification'. 2011. pp. 9-15

[5] Ahtinen A et al. Mobile mental wellness training for stress management: Feasibility and design implications based on a one-month field study. JMIR mHealth and uHealth. 2013;**1**(2):e11

[6] Cafazzo JA, Casselman M, Hamming N, Katzman DK, Palmert MR. Design of an mHealth app for the self-management of adolescent type 1 diabetes: A pilot study. Journal of Medical Internet Research. 2012;**14**(3):e70

[7] Huotari K, Hamari J. Defining gamification: A service marketing perspective. In: Proceeding of the 16th International Academic MindTrek Conference on—MindTrek '12; Tampere, Finland; 2012. pp. 17-22

[8] Alhammad MM, Moreno AM. Gamification in software engineering education: A systematic mapping. Journal of Systems and Software. 2018;**141**:131-150

[9] Dubois DJ, Tamburrelli G. Understanding gamification mechanisms for software development. In: Proceedings of the 2013 9th Joint Meeting on Foundations of Software Engineering—ESEC/FSE 2013; Saint Petersburg, Russia; 2013. pp. 659-662

[10] Sever NS, Sever GN, Kuhzady S. The evaluation of potentials of gamification in tourism marketing communication. International Journal of Academic Research in Business and Social Sciences. 2015;**5**(10):188-202

[11] Almaliki M, Jiang N, Ali R, Dalpiaz F. Gamified culture-aware feedback acquisition. In: 2014 IEEE/ACM 7th International Conference on Utility and Cloud Computing; London, United Kingdom; 2014. pp. 624-625

[12] Lucassen G, Jansen S. Gamification in consumer marketing—Future or fallacy? Procedia - Social and Behavioral Sciences. 2014;**148**:194-202

[13] Yonemura K et al. Effect of security education using KIPS and gamification theory at KOSEN. In: 2018 IEEE Symposium on Computer Applications & Industrial Electronics (ISCAIE); Penang; 2018. pp. 255-258

[14] Yonemura K, Yajima K, Komura R, Sato J, Takeichi Y. Practical security education on operational technology using gamification method. In: 2017 7th IEEE International Conference on Control System, Computing and Engineering (ICCSCE); Penang; 2017. pp. 284-288

[15] Helf C, Zwickl P, Hlavacs H, Reichl P. mHealth stakeholder integration: A gamification-based framework-approach towards behavioural change. In: Proceedings of the 13th International Conference on Advances in Mobile Computing and Multimedia—MoMM 2015; Brussels, Belgium; 2015. pp. 268-274

[16] Nevin CR et al. Gamification as a tool for enhancing graduate medical education. Postgraduate Medical Journal. 2014;**90**(1070):685-693

[17] He Q, Ant AI. A Framework for Modeling Privacy Requirements in Role Engineering. In: Proceedings of

the 9th International Workshop on Requirements Engineering: Foundation for Software Quality (REFSQ'03); 2003

[18] Miyazaki S, Mead N, Zhan J. Computer-aided privacy requirements elicitation technique. In: 2008 IEEE Asia-Pacific Services Computing Conference; Yilan, Taiwan; 2008. pp. 367-372

[19] Rottondi C, Verticale G. Enabling privacy in a gaming framework for smart electricity and water grids. In: 2016 International Workshop on Cyber-physical Systems for Smart Water Networks (CySWater); Vienna, Austria; 2016. pp. 25-30

[20] Shahri A, Hosseini M, Phalp K, Taylor J, Ali R. Towards a code of ethics for gamification at enterprise. In: Frank U, Loucopoulos P, Pastor Ó, Petrounias I, editors. The Practice of Enterprise Modeling. Vol. 197. Berlin, Heidelberg: Springer; 2014. pp. 235-245

[21] Herzig P, Ameling M, Schill A. A generic platform for enterprise gamification. In: 2012 Joint Working IEEE/IFIP Conference on Software Architecture and European Conference on Software Architecture; Helsinki, Finland; 2012. pp. 219-223

[22] Hamari J, Koivisto J, Sarsa H. Does gamification work?—A literature review of empirical studies on gamification. In: 2014 47th Hawaii International Conference on System Sciences; Waikoloa, HI; 2014. pp. 3025-3034

[23] Edwards EA et al. Gamification for health promotion: Systematic review of behaviour change techniques in smartphone apps. BMJ Open. 2016;**6**(10):e012447

[24] Feldbusch L, Winterer F, Gramsch J, Feiten L, Becker B. SMILE goes gaming: Gamification in a classroom response system for academic teaching. In: Proceedings of the 11th International Conference on Computer Supported Education; Heraklion, Crete, Greece; 2019. pp. 268-277

[25] Mavroeidi A-G, Kitsiou A, Kalloniatis C, Gritzalis S. The role of gamification in cultural informatics. In: 2018 Cultural Informatics, Communication and Media Studies (CICMS) Conference; Kusadasi, Turkey. 2018. pp. 43-55

[26] Mavroeidi A-G, Kitsiou A, Kalloniatis C. The interrelation of game elements and privacy requirements for the design of a system: A metamodel. In: Gritzalis S, Weippl ER, Katsikas SK, Anderst-Kotsis G, Tjoa AM, Khalil I, editors. Trust, Privacy and Security in Digital Business. Vol. 11711. Cham: Springer International Publishing; 2019. pp. 110-125

[27] Mavroeidi A-G, Kitsiou A, Kalloniatis C, Gritzalis S. Gamification vs. privacy: Identifying and analysing the major concerns. Future Internet. 2019;**11**(3):67-83

[28] Seaborn K, Fels DI. Gamification in theory and action: A survey. International Journal of Human-Computer Studies. 2015;**74**:14-31

[29] Morford ZH, Witts BN, Killingsworth KJ, Alavosius MP. Gamification: The intersection between behavior analysis and game design technologies. Behavior Analyst. 2014;**37**(1):25-40

[30] Gåsland MM. Game mechanic based E-learning—A case study [MSc. dissertation]. Norway: Norwegian University of Science and Technology; 2011

[31] Werbach K, Hunter D. For the Win: How Game Thinking Can Revolutionize your Business. Philadelphia: Wharton Digital Press; 2012

[32] Merino de Paz B. Gamification: A tool to improve sustainability efforts [Ph.D. dissertation]. England: University of Manchester; 2013

[33] Morschheuser B, Hamari J, Werder K, Abe J. How to gamify? A method for designing gamification. In: Presented at the Hawaii International Conference on System Sciences; 2017. pp. 1298-1307

[34] Chen Y, Pu P. HealthyTogether: Exploring social incentives for mobile fitness applications. In: Proceedings of the Second International Symposium of Chinese CHI on—Chinese CHI '14; Toronto, Ontario, Canada; 2014. pp. 25-34

[35] Cheong C, Cheong F, Filippou J. Quick quiz: A gamified approach for enhancing learning. In: 2013 Pacific Asia Conference on Information Systems; Jeju Island, Korea; 2013. pp. 1-14

[36] Cramer H, Rost M, Holmquist LE. Performing a check-in: Emerging practices, norms and 'conflicts' in location-sharing using foursquare. In: Proceedings of the 13th International Conference on Human Computer Interaction with Mobile Devices and Services—MobileHCI '11; Stockholm, Sweden; 2011. pp. 57-66

[37] McDaniel R, Lindgren R, Friskics J. Using badges for shaping interactions in online learning environments. In: 2012 IEEE International Professional Communication Conference; Orlando, FL, USA; 2012. pp. 1-4

[38] Denny P. The effect of virtual achievements on student engagement. In: Proceedings of the SIGCHI Conference on Human Factors in Computing Systems—CHI '13; Paris, France; 2013. pp. 763-772

[39] Yu-kai Chou: Gamification & Behavioral Design. Available from: http://yukaichou.com/gamification-examples/top-10-marketing-gamification-cases-remember/ [Accessed: 10 December 2019]

[40] Yu-kai Chou: Gamification & Behavioral Design. Available from: https://yukaichou.com/gamification-examples/top-ten-gamification-healthcare-games/ [Accessed: 10 December 2010]

[41] Bista SK, Nepal S, Paris C. Engagement and cooperation in social networks: Do benefits and rewards help? In: 2012 IEEE 11th International Conference on Trust, Security and Privacy in Computing and Communications; Liverpool, United Kingdom; 2012. pp. 1405-1410

[42] Yu-kai Chou: Gamification & Behavioral Design. Available from: https://yukaichou.com/gamification-examples/top-10-education-gamification-examples/ [Accessed: 10 December 2019]

[43] Amo LC, Liao R, Rao HR, Walker G. Effects of leaderboards in games on consumer engagement. In: Proceedings of the 2018 ACM SIGMIS Conference on Computers and People Research—SIGMIS-CPR'18; Buffalo-Niagara Falls, NY, USA; 2018. pp. 58-59

[44] Huotari K, Hamari J. 'Gamification' from the perspective of service marketing. In: ACM Conference on Human Factors in Computing Systems; Vancouver, Canada; 2011

[45] Schmidt-Kraepelin M, Schöbel S, Thiebes S, Sunyaev A. Users' game design element preferences in health behavior change support systems for physical activity: A best-worst-scaling approach. In: 2019 Fortieth International Conference on Information Systems; Munich, Germany. 2019. pp. 1-17

[46] King D, Greaves F, Exeter C, Darzi A. 'Gamification': Influencing

health behaviours with games. Journal of the Royal Society of Medicine. 2013;**106**(3):76-78

[47] Johnson D, Deterding S, Kuhn K-A, Staneva A, Stoyanov S, Hides L. Gamification for health and wellbeing: A systematic review of the literature. Internet Interventions. 2016;**6**:89-106

[48] Xu F, Weber J, Buhalis D. Gamification in tourism. In: Xiang Z, Tussyadiah I, editors. Information and Communication Technologies in Tourism 2014. Cham: Springer International Publishing; 2013. pp. 525-537

[49] Pedreira O, García F, Brisaboa N, Piattini M. Gamification in software engineering—A systematic mapping. Information and Software Technology. 2015;**57**:157-168

[50] Pattakou A, Mavroeidi A-G, Diamantopoulou V, Kalloniatis C, Gritzalis S. Towards the design of usable privacy by design methodologies. In: 2018 IEEE 5th International Workshop on Evolving Security & Privacy Requirements Engineering (ESPRE); Banff, AB; 2018. pp. 1-8

[51] Deng M, Wuyts K, Scandariato R, Preneel B, Joosen W. A privacy threat analysis framework: Supporting the elicitation and fulfillment of privacy requirements. Requirements Engineering. 2011;**16**(1):3-32

[52] Mouratidis H, Shei S, Delaney A. A security requirements modelling language for cloud computing environments. Software and Systems Modeling. 2019;**287**:337-345

[53] Kalloniatis C, Belsis P, Gritzalis S. A soft computing approach for privacy requirements engineering: The PriS framework. Applied Soft Computing. 2011;**11**(7):4341-4348

[54] Islam S, Mouratidis H, Kalloniatis C, Hudic A, Zechner L. Model based process to support security and privacy requirements engineering. International Journal of Secure Software Engineering. 2012;**3**(3):1-22

[55] Liu L, Yu E, Mylopoulos J. Security and privacy requirements analysis within a social setting. In: Journal of Lightwave Technology; Monterey Bay, CA, USA; 2003. pp. 151-161

[56] Pattakou A, Kalloniatis C, Gritzalis S. Security and privacy requirements engineering methods for traditional and cloud-based systems: A review in 2017 cloud computing. In: GRIDs, and Virtualization Conference; 2017. pp. 145-151

[57] Jensen C, Tullio J, Potts C, Mynatt ED. STRAP: A Structured Analysis Framework for Privacy. Georgia Institute of Technology; 2005

[58] Kalloniatis C, Kavakli E, Kontellis E. Pris tool: A case tool for privacy-oriented requirements engineering. In: Doukidis G, et al., editors. Mediterranean Conference on Information Systems; Athens, Greece; 2009

[59] Argyropoulos N, Kalloniatis C, Mouratidis H, Fish A. Incorporating privacy patterns into semi-automatic business process derivation. In: 2016 IEEE Tenth International Conference on Research Challenges in Information Science (RCIS); Grenoble, France; 2016. pp. 1-12

[60] Messner KT. Active-learning simulation-based approach to digital privacy awareness and security in social-media [MSc. dissertation]. 2019

[61] Wilson M, Hash J. Building an information technology security awareness and training program. In: National Institute of Standards and Technology; Gaithersburg, MD; NIST SP 800-50; 2003

[62] Omoronyia I, Cavallaro L, Salehie M, Pasquale L, Nuseibeh B.

Engineering adaptive privacy: On the role of privacy awareness requirements. In: 2013 35th International Conference on Software Engineering (ICSE); San Francisco, CA, USA; 2013. pp. 632-641

[63] Bryce J, Klang M. Young people, disclosure of personal information and online privacy: Control, choice and consequences. Information Security Technical Report. 2009;**14**(3):160-166

[64] Gjertsen EGB. Use of gamification in security awareness and training programs [MSc. dissertation]. 2016

[65] Sheng S et al. Anti-phishing Phil: The design and evaluation of a game that teaches people not to fall for phish. In: Proceedings of the 3rd Symposium on Usable Privacy and Security—SOUPS '07; Pittsburgh, Pennsylvania; 2007. pp. 88-99

Chapter 6

Risk Assessment in IT Infrastructure

Bata Krishna Tripathy

Abstract

Due to large-scale digitization of data and information in various application domains, the evolution of ubiquitous computing platforms and the growth and usage of the Internet, industries are moving towards a new era of technology. With this revolution, the IT infrastructure of industries is rapidly undergoing a continuous change. However, the insecure communication channel; intelligent adversaries in and out of the scene; and loopholes in the software and system development add complexity in deployment of the IT infrastructure in place. In addition, the heterogeneous service level requirements from the customers, service providers, users, along with implementation policies in industries add complexity to this problem. Hence, it is necessary to assess the risk associated with the deployment of the IT infrastructure in industries to ensure the security of the assets involved. In this chapter, we present an efficient risk assessment mechanism in IT infrastructure deployment in industries, which ensures a strong security perimeter over the underlying organizational resources.

Keywords: IT infrastructure, loopholes, service level requirements, common vulnerability scoring system (CVSS), vulnerability, exposure, threat, risk

1. Introduction

In today's world, every industry has their own business goals and functions. In this digital era, industries completely rely on automated information technology (IT) systems to process and manage their typical information to achieve their business objectives. The large-scale digitization of data and information across the various domain, the evolution of ubiquitous computing platforms and growth and usage of the Internet have steered the deployment of information technology systems in industries. IT infrastructure enables efficient service provisioning to end users from various enterprise applications based on Service Level Agreements (SLAs) and dynamic requirements in terms of policies by maintaining the global view of the system. Hence, information technology has become the economic backbone of any industry and offers significant advantages in global markets.

Information technology in an organization includes heterogeneous entities such as general-purpose computing systems, specialized control systems, communication network entities, database management systems, and various software control modules. The integration of these diverse entities helps in the growth and development of an organization by providing reliability, efficiency and robustness of typical information systems as well as business process flow. Despite the advantages provided by the implementation of IT in organizations, open access-control by different levels of users, ubiquitous execution of software modules and control management introduce various security threats. These threats open the door for potential vulnerabilities, environmental interruptions, and inevitable errors leading to different cyber attacks. These attacks can extend to Denial of Service (DoS), code injection, and hidden tunnel, etc. As a result of various attacks, the confidentiality, integrity, availability (CIA) of the critical information is severely compromised. This, in turn, may have a huge impact on organizational assets, business operations, individuals, other stakeholders, and above all the Nation's assets.

Researchers have witnessed that as compared to outside threats there are pre-eminent threats from inside users and entities in organizations [1]. The organizations must understand the importance and responsibilities for protecting critical organizational information, assets, and processes from intelligent attackers. It has become an imperative duty of the organization for assessing the risk associated with the operation and use of different entities in information technology systems. Risk assessment is a key discipline for making effective business decisions by identifying potential managerial and technical problems in IT infrastructure. Then, necessary remediation can be taken by the managers of the organization to minimize or eliminate the probability and impact of these problems.

This chapter presents an efficient risk assessment mechanism that proactively analyzes the risks of IT infrastructure creating strong isolation between different entities. The proposed risk assessment solution determines the threat associated with different entities by analyzing vulnerability and exposure with respect to the Common Vulnerability Scoring System (CVSS) [2]. The overall risk of the IT systems is calculated as the cumulative threat values of different entities. These risk measures, in turn, drive the remediation process for appropriate risk mitigation in the organization strengthening the security perimeter of the organizational resources.

The rest of the chapter is organized as follows. Section 2 presents the related works in risk assessment in IT infrastructure. Section 3 presents the background of the risk assessment of IT infrastructure in organizations. The steps of risk assessment are discussed in Section 4. Section 5 presents our proposed IT risk assessment framework in detail. Section 6 summarizes the chapter.

2. Related works

Security risk assessment in enterprise networks has ever remained a major challenge for research communities. Defining security metrics play an important role in risk assessment. The literatures [3–5] define various security metrics. The effectiveness of a risk assessment mechanism relies on the security metric considered during the risk evaluation process.

The primitive risk management mechanisms were qualitative-based which used the System Security Engineering-Capability Maturity Model (SSE-CMM) using attack graphs [6]. However, these works do not evaluate risk quantitatively which can play a major role in identifying several threats. Later, the Common Vulnerability Scoring System (CVSS) [2] was proposed which is used for quantitative risk evaluation. VRSS [7] is another quantitative approach that evaluates risk using varieties of vulnerability rating systems. This uses statistics from different vulnerability databases such as IBM ISS X-Force, Vupen Security, and National Vulnerability database to determine overall risk measure in an organization. However, these works significantly lack accurate evaluation of risk in an enterprise network because of the security metrics considered and the evaluation process.

The work [4] presents a quantitative risk assessment method that determines the threat value from the number of attacks in a specific time interval. Munir et al. [8] proposed another quantitative risk assessment method using the vulnerability scanning tool (Nexpose) to determine the vulnerability values in each node in the network. This method uses the CVSS and the probabilistic approach to determine an overall risk measure of the enterprise network. In another work [9] the risk of the network is analyzed by determining the impact and likelihood of vulnerabilities. It uses WPA2 as the basic cryptographic algorithm.

On the other hand, Guohua [10] presented a risk assessment technique based on AHP (Analytic Hierarchy Process) which quantitatively determines the confidentiality, integrity, and availability of the assets with respect to the individual asset classes. In another work, Munir et al. [11] proposed a risk assessment mechanism based on the classification of different attacks as per their characteristics. This work also implements a method using a rule in Snort NIDPS signature database and OWASP risk rating approach to determine the overall risk of an enterprise network.

In a recent work, Lamichhane et al. [12] presented a quantitative risk assessment approach which computes risk as a function of overall vulnerabilities exploitation along a path and impact of the exploitation. This work implements Topological Vulnerability Analysis (TVA) for modeling and analysis of attack paths using attack graph. Chalvatzis et al. [13] proposed a virtual machine based testing framework for the performance of vulnerability scanners of the enterprise networks. The literature presented a comparative statistics of the vulnerability scanning solutions such as Nessus, OpenVAS, Nmap Scripting Engine with respect to their automation risk assessment process.

However, the state of art works do not accurately determine the risk of the enterprise network considering the risk associated with individual assets, the impact, and criticality of the information flow. In this chapter, we present an efficient risk assessment mechanism in IT infrastructure deployment in industries which addresses the limitations of the existing risk assessment techniques. Our proposed solution ensures a strong security perimeter over the underlying organizational resources by considering the level of vulnerability, threat, and impact at individual assets as well as the criticality of the information flow in the organization.

3. Background

The managers and stakeholders of organizations must understand and identify the different parameters necessary for assessing the risk of IT infrastructure. These parameters are defined as follows.

3.1 Vulnerability

It is defined as a software and hardware level weakness in the entities of IT systems, which may allow an attacker to reduce the information assurance of the entities and the underlying network [14]. In other words, it is the source of a known problem that opens the door for a potential attack on the IT infrastructure system. For example, if the managers of an organization mistakenly do not disable the access to resources and processes such as logins to internal systems for an ex-employee, then this leads to both unexpected threats to the IT infrastructure. In most cases, the vulnerabilities are exploited intentionally or unintentionally by inside or outside users of the IT systems and have a severe impact on the organizational assets. Hence, identifying weak points in the entities of IT systems is the first step to managing the risk of the IT infrastructure to ensure reliability, robustness, efficiency, and security of IT resources.

3.2 Exposure

It is defined as the state or condition of a system being unprotected and open to the risk of suffering the loss of information [15]. In general, exposure of an entity may be a malicious piece of code, commands, or open-source tools that may potentially cause system configuration issues. This, in turn, may allow attackers to track business process flow as well as to gather critical information and at far can lead to gain access to even whole IT infrastructure. Determining exposure is the primary objective of an attacker for discovering a vulnerability in the IT systems. Generally, the exposure of an entity in the IT systems is represented as the ratio of the potentially unprotected portion of the entity to the total entity size.

3.3 Threat

Threats are potential events for vulnerabilities that might lead to exposure of the network and adversely impact the organizational assets [16]. A threat has the potential of causing small to even severe damage to the IT infrastructure of organizations. The source or root of threats can be natural, intentional or unintentional. Natural threats can be catastrophe such as floods, cyclones, earthquakes, etc. On the other hand, unintentional threats can be mistakes done by employees of organizations such as accessing the wrong resources. Intentional threats are created by attackers by flooding malicious codes over the network in the form of spyware, malware, worms, viruses, etc. Most recently, on Oct 24, 2019, Ransomware and DDoS attacks brought down major banks in South Africa including Johannesburg demanding a ransom of four Bitcoins that is equivalent to about R500,000 South African Rand or $37,000 USD [17]. Vulnerability and exposure of an entity are used to determine its threat value.

3.4 Risk

It is defined as an uncertain incident created as a result of a system malfunction and in turn has a severe impact on organizational assets and business objectives [18]. In general, the risk is a qualitative measure of potential security threat and its impact on the network [19]. In other words, the risk is defined as the potential for harm to organizations' resources when a vulnerability is exploited to threat. For example, the risk may include loss of privacy, financial loss, legal complications, etc. Hence, the overall risk of the IT systems is assessed by analyzing the vulnerability, exposure, and threat of different entities in the IT infrastructure.

Risk assessment plays a key role in making and implementing effective business decisions by proactively identifying potential problems at different managerial and technical levels. Risk management, therefore, can follow necessary remediation steps to overcome the severity of these problems [20].

4. Steps for IT risk assessment

An effective IT risk assessment process in an organization comprises the following major steps or phases. These steps are similar to the steps illustrated in the work [21]. However, we have considered the sub-phases of the evaluation phase, that is, identifying vulnerabilities, determining exposure, determining threat as different phases in our work since these steps are equally important as compared to other phases. The risk assessment process follows a life cycle with these steps or phases as shown in **Figure 1** aiming to eliminate or minimize the level of risks in the IT infrastructure.

4.1 Step 1: Evaluation

In this phase, the critical resources that may have potential vulnerability and have threats must be understood and identified. The critical resources include the process flows, enterprise information, and assets in the IT infrastructure that are important for the functioning and security of the business. This, in turn, helps in understanding the consequences of critical information loss and in decision making regarding the resources that need to be protected.

4.2 Step 2: Identifying vulnerabilities

In this phase, the inherent vulnerabilities in the entities of IT systems are reviewed, identified and listed that have potential threats to affect the organizational assets and business process. This includes both software and hardware-level vulnerabilities of IT infrastructure. The list of vulnerabilities must have detailed information such as type, impact, measure, etc.

4.3 Step 3: Determining exposure

In this phase, the exposure of the entities in the IT systems that may have a potential threat to different attacks is determined and reported. Generally, the exposure of an entity in the IT systems is computed as the ratio of the potentially unprotected portion of the entity to the total entity size.

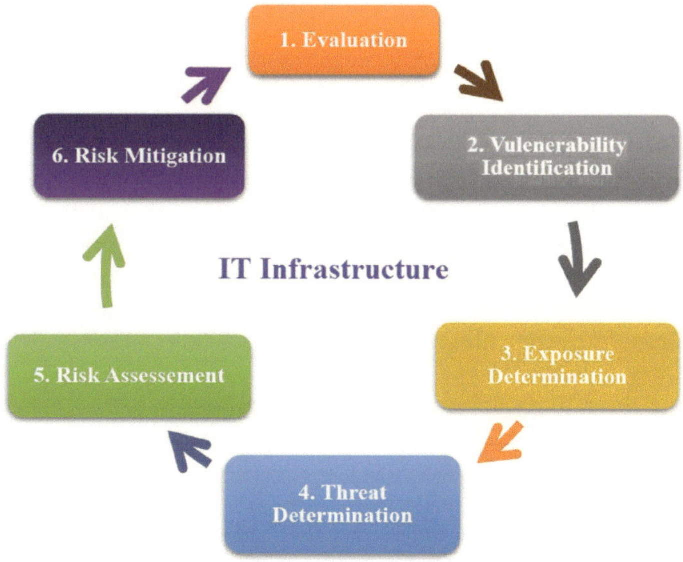

Figure 1.
Risk assessment life cycle in IT infrastructure.

4.4 Step 4: Determining threat

In this phase, information on potential threats to the organizational assets and information is gathered that may have a direct or an indirect impact on the business process. This includes collecting details of the threats on each IT entities from inside and outside users or attackers. This ultimately guides the risk assessment process for the necessary remediation plan and action to protect the organizational resources.

4.5 Step 5: Risk assessment

This phase focuses on determining the probability and impact of the vulnerabilities in the entities of IT systems. As all threats do not have the likelihood of equal occurrence and impact on the organizations' infrastructure, so it is crucial to correctly identify different levels of risk. Hence, each level of risk is determined by mapping individual threats, exposure, and vulnerabilities of an entity based on their probability and impact to critical resources of the organization. This, in turn, helps in decision making on the implementation of appropriate remediation acts.

4.6 Step 6: Risk mitigation

Once the risk assessment is performed, the final step for IT managers is to plan and act according to take preventive measures for potential threats to the organizations. It may consist of different measures such as identifying different threats before their occurrence, minimizing or eliminating the consequences of security breaches, recovering to a safe state to resume normal business process, etc.

5. IT risk assessment framework

In this chapter, an effective IT assessment framework is presented to ensure a strong security perimeter over the vulnerable IT environment of the organizations.

5.1 Vulnerability analysis

The Common Vulnerability Scoring System (CVSS) [2] plays an important role in the risk assessment of the entities in the IT infrastructure to ensure secure business information flow across the IT systems. The risk assessment module uses a data structure called vulnerability database for this purpose. The vulnerability database is a local repository (offline) stored in the controller. It is periodically updated with the recent Common Vulnerability Score (CVS) values of the applications or protocols or services running in different hardware and software components or entities of IT infrastructure. The CVS values are computed by extracting necessary metrics from the online National Vulnerability Database (NVD) [22] using a script.

The recent vulnerability values available in NVD are in XML format which contains two standard scores: V2 and V3 in the form of Common Vulnerability and Exposure (CVE) measures. The detailed process of parsing CVE values from NVD and storing in the local vulnerability database as CVS values is explained in **Figure 2**. It is to be noted that in the vulnerability database, there exists exactly one entry of CVS value for an application with its version and the Operating System platform as it is the updated CVSS value of the application parsed from NVD's recent XML file using the script. The structure of an entry in the vulnerability database is <*Application/service/protocol*, *Version*, *Operating system*, *CVS value*>.

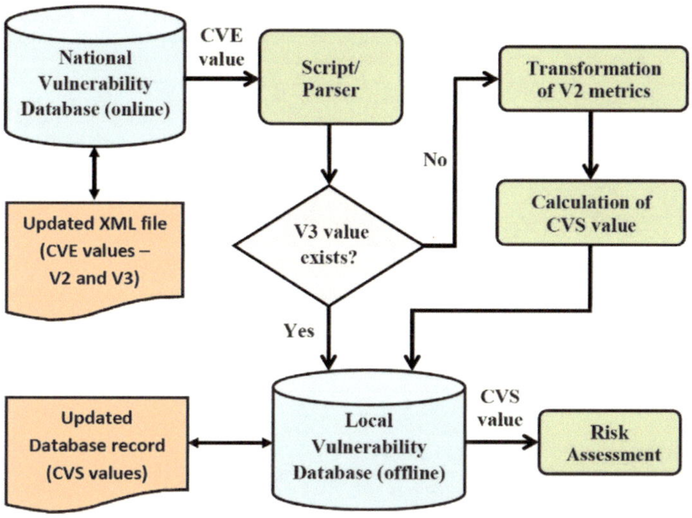

Figure 2.
Parsing CVE values from NVD and storing as CVS values in local vulnerability database for risk assessment.

Generally, the V3 standard is an improvement over the V2 standard as V3 considers the context of attacker's access rights to read/write/execute to exploit the vulnerability and physical manipulation of the affected components. Hence, the risk assessment module uses the V3 version of CVE as its CVS value for necessary risk assessment for secure business processes and information flow. However, for some older vulnerabilities there exist only V2 values in NVD. In such a case, the CVS value for a vulnerability is calculated in two steps from the available V2 metrics in NVD as discussed below.

5.1.1 Step 1: Transformation of V2 metrics

To compute the overall vulnerability value, CVSS considers certain metrics that define the hardware, software and network-level vulnerabilities in the IT systems. The V2 version differs from the V3 version in terms of the metrics and their values considered for overall vulnerability score computation. However, for some older vulnerabilities, V3 value is not available in the NVD. In this scenario, the CVS value for a vulnerability in our solution is estimated from the V2 metrics available in the XML file by appropriately transforming the metrics and their values as shown in **Table 1**. The transformation is performed as per the CVSS V2 and V3 standards [23, 24].

These metrics after the transformation process are then used for the necessary CVS computation in the proposed mechanism. The estimation of CVS value for a vulnerability is performed as explained below in the subsequent step.

5.1.2 Step 2: Calculation of CVS values

The CVS value for a vulnerability is determined from the desired metrics obtained in the previous step, using the standard equations for the overall V3 version of CVSS computation [24] with optimization to minimize the overhead of the CVS computation process. The procedure of the overall CVS value calculation is illustrated in **Figure 3**.

The entities of the IT infrastructure might be the potential sources of vulnerabilities in the organization. Hence, the vulnerability of each entity is determined by

V2 metric		**Value**	**Transformed metric**	**Value**
Base metrics				
Exploitability group	Access vector	Local: 0.395	Attack vector	Local: 0.55
		Adjacent network: 0.646		Adjacent: 0.62
		Network: 1.0		Network: 0.85
	Access complexity	High: 0.35	Attack complexity	High: 0.44
		Medium: 0.61		Medium: 0.62
		Low: 0.71		Low: 0.77
	Authentication	Multiple: 0.45	Privileges required	High: 0.27
		Single: 0.56		Low: 0.62
		None: 0.704		None: 0.85
Impact group	Confidentiality, integrity, and availability	None: 0.0	Confidentiality, integrity, and availability	None: 0.0
		Partial: 0.275		Low: 0.22
		Complete: 0.66		High: 0.56
Temporal metrics				
	Exploitability	Unproven: 0.85	Exploitability	Unproven: 0.91
		Proof-of-concept: 0.9		Proof-of-concept: 0.94
		Functional: 0.95		Functional: 0.97
		High: 1.0		High: 1.0
	Remediation level	Official fix: 0.87	Remediation level	Official fix: 0.95
		Temporary fix: 0.90		Temporary fix: 0.96
		Workaround: 0.95		Workaround: 0.97
		Unavailable: 1.0		Unavailable: 1.0
	Report confidence	Unconfirmed: 0.90	Report confidence	Unknown: 0.92
		Uncorroborated: 0.95		Reasonable: 0.96
		Confirmed: 1.0		Confirmed: 1.0
Environmental metrics				
General modifiers	Collateral damage potential	None: 0	Attack vector	None: 0
		Low (light loss): 0.1		Physical: 0.2
		Low-medium: 0.3		Local: 0.55
		Medium-high: 0.4		Adjacent network: 0.62
		High (catastrophic loss): 0.5		Network: 0.85
	Target distribution	None: 0	Attack complexity	None: 0
		Low: 0.25		Low: 0.77
		Medium: 0.75		Medium: 0.62
		High: 1.0		High: 0.44

V2 metric		Value	Transformed metric	Value
Impact subscore modifier	Confidentiality, integrity, and availability requirements	Low: 0.5	Confidentiality, integrity, and availability requirements	Low: 0.5
		Medium: 1.0		Medium: 1.0
		High: 1.51		High: 1.5

Table 1.
Transformation of V2 metrics and their values for CVS computation.

the above-mentioned steps. Then, the threat for different entities is determined using the threat model using vulnerability and exposure analysis of those entities. Then, the overall risk of the IT systems is determined as cumulative threat values of the entities and criticality of the business process and information flow.

5.2 Threat model

In this phase, the threat associated with different IT entities is modeled using the vulnerability and exposure of the entities as follows.

5.2.1 Vulnerability of an entity

Several vulnerable applications, services or protocols such as FTP, RSH, Nmap, etc. may be running in an IT entity for the functioning of business processes. The vulnerability V_e of an entity e is calculated as the average of the Common Vulnerability Scores (CVS) of all the applications running on the entity extracted from the vulnerability database, that is,

$$V_e = \frac{1}{10} * \frac{\sum_{i=1}^{k} CVS_i}{k} \quad (1)$$

where CVS_i is the Common Vulnerability Score of the ith application or protocol or service running in the entity e, and k is the number of applications, protocols, and/or services running in the entity. The average value of the CVS of all applications, protocols and/or services is divided by 10 to normalize the value of V_e to 1 as the CVS lies between 0 and 10.

5.2.2 Exposure of an entity

The exposure E_e of an entity e is determined considering the number of entities that may be affected because of the vulnerability in the target entity. Hence, it is computed as,

$$E_e = \frac{n}{N} \quad (2)$$

where n is the number of entities communicating with the target entity and N is the total number of entities in the IT systems.

The vulnerability values and threat models guide the risk assessment process for estimating risk levels of the entities in the IT infrastructure.

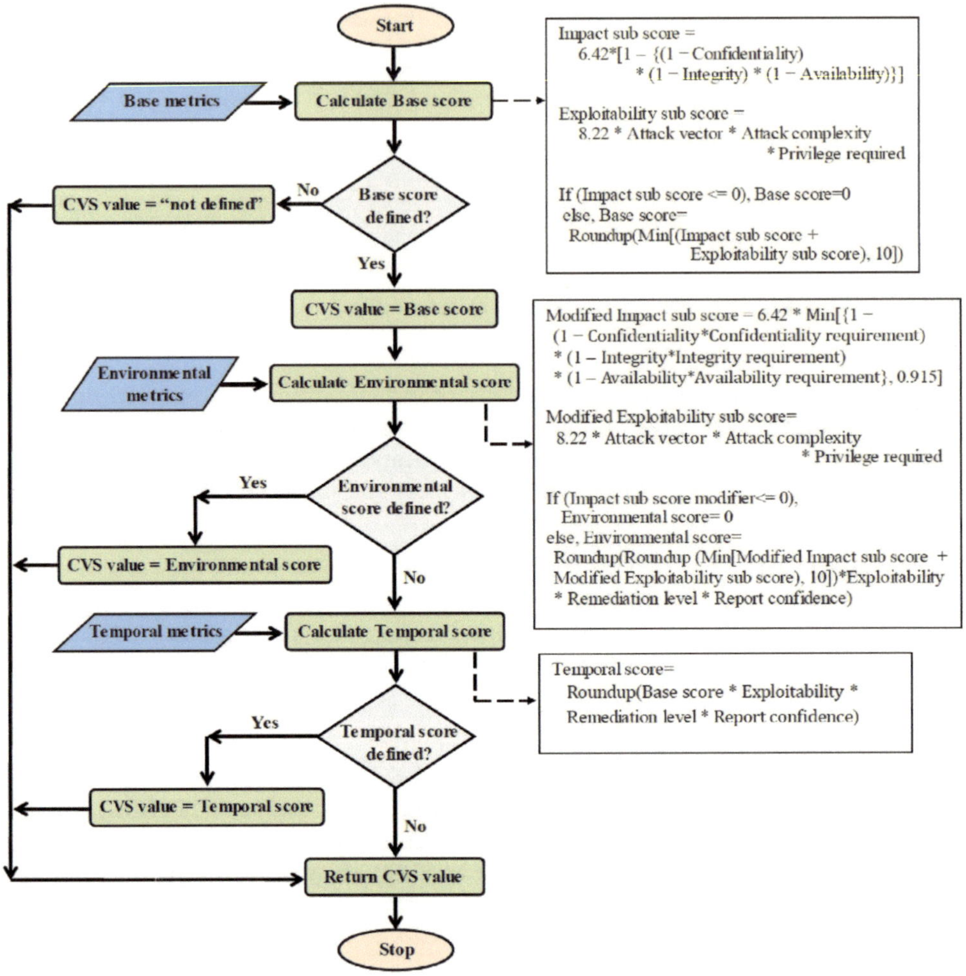

Figure 3.
CVS computation of vulnerabilities from the transformed metrics in case of nonavailability of V3 value in NVD.

5.3 Risk assessment model

The risk assessment model first evaluates the threat model for different IT entities as discussed in the previous subsection. Then, the overall threat value τ is calculated as the cumulative threat values of all the entities in the IT systems involved in the business process flow. Algorithm 1 illustrates the risk assessment procedure to determine the overall threat value τ.

Algorithm 1 uses weight w_e for each entity in order to consider the criticality of different entities and should be chosen such that their sum must be equal to 1, that is,

$$\sum w_e = 1 \tag{3}$$

In our work, we have used the term *weight* as it is a quantitative term instead of the term *criticality* which is usually a qualitative term.

The overall threat value (τ) and its criticality (I) of business process and information flow are used to define the overall risk (R) of the entities in IT systems. The criticality of the business process and information flow can be high (H),

Algorithm 1 *Risk assessment* algorithm

procedure RISK_ANALYZE
    Entity set $E = e_1, e_2, \ldots, e_n$
    for each entity $e \in E$ **do**
        find V_e
        find E_e
        calculate $\tau_e = V_e{}^{*}E_e$
    end for
    calculate $\tau = \sum w_e{}^{*}\tau_e$
end procedure

medium (M) or low (L). The criticality of a business process and information flow depends on the impact of the business process and information flow in a specific application context. For example, in a banking application, transactions have high impact and hence have High importance whereas the generation of logs has medium impact leading to medium importance. On the other hand, simple query processing has a low impact on the context and hence has low importance. So, we consider three different criticality levels; that is, high (H), medium (M) and low (L), respectively for these three types of business process and information flow.

The mapping function for assessing the risk of a specific business process and information flow is expressed as:

$$f : \tau \times I \rightarrow R \tag{4}$$

Table 2 shows the risk assessment model of IT infrastructure with respect to the criticality and threat level of the specific business process and information flow in the enterprise network. For example, in a banking application, transactions have high impact and hence have high criticality whereas the generation of logs has medium impact leading to medium criticality. On the other hand, simple query processing has a low impact on the context and hence has low criticality. So, we consider three different criticality levels of the business process and information flow; that is, high (H), medium (M) and low (L), respectively for overall risk assessment. For example, if the criticality of a business process and information flow is high (H) and its threat value is 5.5, then the risk associated with the business process and information flow is high (H). Similarly, individual risk levels are determined concerning specific business processes and information flow.

The calculated risk measures determined by the risk assessment model, are used in decision making and remediation planning for protecting the systems against different potential attacks. This process is executed recursively to eliminate or minimize the level of risks in the IT infrastructure.

Criticality of business process and information flow	**Total threat value**		
	⩽0.39	**0.4 to 0.69**	**≥0.7**
H	M	H	C
M	L	M	H
L	L	L	M

Note: C, critical; H, high; M, medium; and L, low.

Table 2.
Risk assessment model of IT infrastructure.

6. Conclusion

The evolution of ubiquitous computing systems has steered the industries towards relying on IT infrastructure for their business operations. In addition, industries are competing in the global market adapting to the rapid and continuous changes in IT systems. However, deployment of the IT infrastructure across industries has always remain complicated because of the insecure communication channel; intelligent inside and outside attackers; and loopholes in the software and system development life cycle. In addition, the heterogeneous service level requirements from the customers, service providers, users, along with implementation policies in industries add complexity to this problem. Hence, effective assessment of risk associated with the deployment of the IT infrastructure in industries has become an integral part of the management to ensure the security of the assets. In this chapter, an efficient risk assessment mechanism for IT infrastructure deployment in industries is proposed which ensures a strong security perimeter over the underlying organizational resources by analyzing the vulnerability, threat, and exposure of the entities in the system.

Abbreviations

IT	information technology
NVD	National Vulnerability Database
CVSS	Common Vulnerability Scoring System
CVS	Common Vulnerability Score
CVE	Common Vulnerability and Exposure

Author details

Bata Krishna Tripathy
Indian Institute of Technology Bhubaneswar, India

*Address all correspondence to: bata.krishna.tripathy@gmail.com

References

[1] Insider vs. Outsider Data Security Threats: What's the Greater Risk? [Online]. Available from: https://digitalguardian.com/blog/insider-outsider-data-security-threats. [Accessed: 01 November 2019]

[2] Mell P, Scarfone K, Romanosky S. Common vulnerability scoring system. IEEE Security and Privacy. 2006;**4**(6): 85-89. [Accessed: 01 November 2019]

[3] Li J, Wang H. A quantification method for network security situational awareness based on conditional random fields. In: Fourth International Conference on Computer Sciences and Convergence Information Technology; Seoul; 2009. pp. 993-998. [Accessed: 01 December 2019]

[4] Breu R, Innerhofer-Oberperfler F, Yautsiukhin A. Quantitative assessment of enterprise security system. In: Third IEEE International Conference on Availability, Reliability and Security; Barcelona; 2008. pp. 921-928. [Accessed: 01 December 2019]

[5] Xie A et al. An adjacency matrixes-based model for network security analysis. In: IEEE International Conference on Communications; Cape Town, South Africa; 2010. pp. 1-5. [Accessed: 01 December 2019]

[6] Noel S et al. Measuring security risk of networks using attack graphs. International Journal of Next Generation Computing. 2010;**1**(1): 135-147. [Accessed: 01 December 2019]

[7] Liu Q, Zhang Y. VRSS: A new system for rating and scoring vulnerabilities. Computer Communications. 2011;**34**: 264-273. [Accessed: 01 December 2019]

[8] Munir R, et al. A quantitative measure of the security risk level of enterprise networks. In: Eighth IEEE International Conference on Broadband and Wireless Computing, Communication and Applications; Compiegne; 2013. pp. 437-442. [Accessed: 01 December 2019]

[9] Liang L, et al. The practical risk assessment for enterprise Wireless Local Area Network. In: IEEE International Conference on Information Science, Electronics and Electrical Engineering; Sapporo; 2014. pp. 1936-1940. [Accessed: 01 December 2019]

[10] Guohua Z. Enterprise information security risk and countermeasure research under network environment. In: Seventh IEEE International Conference on Measuring Technology and Mechatronics Automation; Nanchang; 2015. pp. 453-456. [Accessed: 01 December 2019]

[11] Munir R, et al. Detection, mitigation and quantitative security risk assessment of invisible attacks at enterprise network. In: 3rd IEEE International Conference on Future Internet of Things and Cloud; Rome; 2015. pp. 256-263. [Accessed: 01 December 2019]

[12] Lamichhane PB, Hong L, Shetty S. A quantitative risk analysis model and simulation of enterprise networks. In: 9th IEEE Annual Information Technology, Electronics and Mobile Communication Conference (IEMCON); Vancouver, BC; 2018. pp. 844-850. [Accessed: 01 December 2019]

[13] Chalvatzis I, Karras DA, Papademetriou RC. Evaluation of security vulnerability scanners for small and medium enterprises business networks resilience towards risk assessment. In: IEEE International Conference on Artificial Intelligence and Computer Applications (ICAICA); Dalian, China; 2019. pp. 52-58. [Accessed: 01 December 2019]

[14] Vulnerability (Computing), Wikipedia [Online]. Available from: https://en.wikipedia.org/wiki/Vulnerability_(computing). [Accessed: 01 November 2019]

[15] Exposure [Online]. Available from: http://www.businessdictionary.com/definition/exposure.html. [Accessed: 01 November 2019]

[16] Threat (Computer), Wikipedia [Online]. Available from: https://en.wikipedia.org/wiki/Threat_(computer). [Accessed: 01 November 2019]

[17] Cyber Attacks Hit the City of Johannesburg and South African Banks [Online]. Available from: https://www.thesslstore.com/blog/cyber-attacks-hit-the-city-of-johannesburg-and-south-african-banks/. [Accessed: 01 November 2019]

[18] Harvey J, Technical Information Service. Introduction to Managing Risk [Online]. Available from: http://www.cimaglobal.com/Documents/Imported Documents/cid_tg_intro _to_managing_rist.apr07.pdf. [Accessed: 01 November 2019]

[19] Cybersecurity Risk: A Thorough Definition, Bitsight [Online]. Available from: https://www.bitsighttech.com/blog/cybersecurity-risk-thorough-definition. [Accessed: 01 November 2019]

[20] Georgieva K, Farooq A, Dumke RR. Analysis of the risk assessment methods —A survey. In: Software Process and Product Measurement. IWSM 2009. Lecture Notes in Computer Science, Vol. 5891. Berlin, Heidelberg: Springer

[21] IT Risk Assessment—Happiest Minds [Online]. Available from: https://www.happiestminds.com/whitepapers/IT-risk-assessment. [Accessed: 01 November 2019]

[22] National Vulnerability Database (NVD) [Online]. Available from: https://nvd.nist.gov/ [Accessed: 01 November 2019]

[23] A Complete Guide to the Common Vulnerability Scoring System Version 2.0 [Online]. Available from: https://www.first.org/cvss/v2/guide. [Accessed: 01 November 2019]

[24] Common Vulnerability Scoring System v3.0: Specification Document [Online]. Available from: https://www.first.org/cvss/specification-document. [Accessed: 01 November 2019]

Chapter 7

Risks of Privacy-Enhancing Technologies: Complexity and Implications of Differential Privacy in the Context of Cybercrime

William Stadler

Abstract

In recent years, the swift expansion of technology-enabled data harvesting has infiltrated modern life and led to the collection of massive amounts of private data. As a result, the preservation of individual privacy has become a salient concern for the general public. Combined with an increase in the frequency and prevalence of cybercrime, more of the public now face the very real risk of privacy loss associated with illegitimate use of private data. Differential Privacy has emerged as a relatively new privacy-preserving method with the potential to significantly reduce the likelihood of harmful data disclosures stemming from malicious use. However, research has not explicitly investigated Differential Privacy from the perspective of criminal justice or examined the utility of Differential Privacy as a possible situational crime prevention measure to cybercrime. Therefore, this chapter explores the proliferation of cybercrime through advances in technology and briefly examines other privacy-preserving methods before discussing the possible use of Differential Privacy as a viable countermeasure to cybercrime. The chapter concludes with a discussion of several practical considerations related to the use of Differential Privacy as a tool in the fight against cybercrime and offers recommendations for future research.

Keywords: cybercrime, Differential Privacy, privacy-enhancing technologies, technology-enabled crime, situational crime prevention

1. Introduction

The production and consumption of data has been increasing with the ubiquity of the internet [1], and with this, the benefits that accompany innovations and advances in computing technology, such as those stemming from artificial intelligence and machine learning, are increasingly relevant to a growing number of industries and applications [2]. However, our reliance on technology and consumer connectedness, coupled with rapid growth in the aggregation and liquidity of personalized data, has made us more vulnerable to cybercrime victimization and the malicious use of private data [3, 4]. The challenge of securing confidential information is becoming one of the key issues in our digital world [5].

Recently developed privacy-enhancing technologies and methods are being touted as possible solutions to mitigate privacy risks associated with inadvertent disclosure and guard against sinister data incursions resulting from cybercrime. One such possibility is Differential Privacy [6], which represents a new security paradigm designed to meet the growing number of privacy risks which accompany data stewardship, particularly for those entrusted with safeguarding data. Differential Privacy was conceived to simultaneously harness the power of information contained in "big data" while substantially reducing the likelihood of harmful data disclosures resulting in possible malicious use [7].

The commercial benefits and costs of privacy enhancing technologies have been widely studied, particularly as consumer data sharing and consumption has grown through distributed systems and Internet of Things (IOT) devices and applications such as smartphones, televisions, medical equipment, appliances, and wearables. However, because of its emergence as a promising new approach to computational analysis, far less has been written about the implications of Differential Privacy, including the merits and limitations of the sophisticated techniques created in the context of this definition. Similarly, research aimed at the advantages, pitfalls, and practical challenges of adopting differentially private approaches has been limited. Literature on Differential Privacy has yet to explore the applied use of this privacy-preserving approach in the context of contemporary crime and justice threats, including cybercrime. Scholarship has generally tended to avoid important, and arguably necessary, cross-disciplinary collaborations between technical science disciplines such as computer science and social science disciplines like criminal justice.

Therefore, through the lens of the criminal justice discipline, this chapter will explore the use of Differential Privacy as a possible cybercrime prevention technique in the context of the massive digital ecosystem that has emerged over the last two decades. We begin with a discussion of the recent proliferation of cybercrime that has arisen through advances in technology, followed by a brief examination of evolving privacy protections which led to the rise of differential privacy, as both a general tenet and assortment of techniques for advancing data security. We then speculate on the use of Differential Privacy as a situational crime prevention countermeasure to cybercrime, and review potential challenges to its use. The chapter concludes with an attempt to stimulate future research and interest in cross-disciplinary exploration of this relatively new privacy-enhancing approach, particularly with respect to its potential to reduce risk, combat crime, and preserve the confidentiality of data for consumers and those most vulnerable to cybercrime victimization.

2. The proliferation of cybercrime through technology

As the general public engages more with online environments and participation in connected routines that produce personal data becomes more common to everyday life, new criminal opportunities emerge in the form of cybercrime [8]. Though the concept of cybercrime is open to interpretation and has resulted in several competing definitions, broadly defined, cybercrime involves technology-related offending that takes place in the online environment [9] and is "committed using a computer, network, or hardware device" [10]. More importantly, cybercrime represents a serious economic and national security threat to the United States and to other countries around the world [11, 12]. Research has revealed that theft of private data through cybercrime is continuing to grow [1], resulting in a substantial need for promising new definitions and approaches, as well as new laws [13], aimed at the protection of personal data and individual privacy. Differential Privacy is one of many approaches with the potential to prevent or significantly blunt the harmful consequences associated with cyber criminality by influencing the means through which organizations and agencies protect sensitive information from exploitation and malicious use. Yet, cybercrime itself has largely remained on the periphery of the criminology discipline as a marginalized topic [3], and research on information security in the context of cybercrime has remained limited as a result, perhaps because of the complexities associated with the crimes and spatial and temporal distance between offenders and victims [13]. Further, research in crime and justice literature on both the theoretical and practical use of technical privacy methods, such as Differential Privacy, is virtually non-existent.

Meanwhile, the spread of data-driven technologies are generating a multitude of ways for public and private-sector entities to induce the creation and dissemination of personal data which also inadvertently enables cybercriminals access to information that people would rather keep to themselves. Ironically, the recent trend toward distributed computing and the decentralization of control and access to smaller computer systems and network resources has also increased the likelihood of cybercrime [14]. Once data have been generated and exist somewhere, the malicious use of that data becomes more likely, creating greater potential for victimization and harm to individuals and to organizations alike. Thus, two related issues become tantamount when considering the practical utility of Differential Privacy as one of many possible countermeasures to cybercrime. First, it is important to understand how cybercrime threats are evolving and expanding to ensure that subsequent prevention and interdiction measures are designed with specific cybercrime threats in mind. Second, it is also necessary to consider how detection and attribution capabilities have evolved in relation to the changing threat landscape so that cybercrime enforcement and investigation methods also meet changing demands.

2.1 Threat expansion and evolution

The world community has been increasingly expressing concern about the use of advanced computing and AI for criminal purposes [15]. And in recent years, advances in technology have undoubtedly increased the frequency and prevalence of cybercrime activity, resulting in an expansion of possible threats to systems and data worldwide [13]. Given the breadth of information captured and widely available today about each individual on earth, people might assume that the magnitude of the internet and related "systems" as well as volume of data being transmitted provides adequate protection against disclosures of personal data. Individuals sharing this view may also conclude that the odds of becoming a victim are low and that more robust technical countermeasures to cybercrime are unnecessary. However, this perception is a fallacy; vulnerability to victimization is not uniformly distributed, nor are contemporary acts of cybercrime targeted only at single persons or entities. The size and scale of cybercrime capabilities and efforts has increased commensurate with advances to computing power and precision, perhaps resulting in modern cyber-predators posing greater risk to larger groups of individuals than ever before [15].

This fact is becoming more evident as the United States and other countries around the world grapple with increasingly serious cases of cybercrime which strain the integrity of data protection measures in both public and private sectors. Dozens of high-profile and illegal data breaches have occurred in the U.S. over the last handful of years that resulted in the compromise or theft of massive amounts of private information, including with eBay [16], JP Morgan Chase [17], Sony [18], Adobe, Equifax, and LinkedIn [19], as well as with U.S. political organizations [20] and voter registration records [21]. Highly sophisticated gangs, organized crime groups, and terrorist organizations are also using computer and communication technologies to steal, smuggle, blackmail, sell drugs, and conduct a variety of other criminal activities on a much larger scale to finance their operations [22]. To be sure, cybercriminals are becoming more knowledgeable and skilled, and they appear to be systematically attacking larger and more sensitive databases with increasing brazenness and alarming frequency.

Recent advances in privacy technology have to some degree equipped data guardians with more tools to systematically prevent inadvertent data disclosures resulting from legitimate use. With respect to cybercrime, the contribution of new innovations has also enabled private corporations and government agencies, including those serving prevention, enforcement, or regulatory functions, to better deter, investigate, and detect instances of nefarious activity and cybercrime attacks resulting in privacy fissures. Yet, on the whole, governments and private entities frequently appear to be playing catch-up. Growth of distributed systems, AI, and novel privacy enhancing technologies which strengthen the capabilities of data producers and distributors have also produced unintended consequences, including conditions favorable to hostile actors gaining the motivation, means, and cover to access private information and conceal malicious activity [23]. Moreover, typical privacy protections have achieved limited success because they are inattentive to the opportunistic aspects of cybercrime [14]. Commonly deployed data protection tactics may generate a false sense of security while inadvertently softening crime targets by making them more attractive, accessible, and unguarded to allow cybercriminals opportunities to conceivably initiate attacks on private information more easily. The resulting "target softening" stems directly from the shift toward complex software, interconnected data networks, and distributed systems in the modern IoT infrastructure which remain inadequately guarded and vulnerable to penetration via more sophisticated techniques [5]. While innovations and capabilities advancements undoubtedly enable more sophisticated applications, they also enable adversaries to collect information and deliver exploits specifically tailored to target systems [24].

The frequency of hostile attacks will also likely increase as artificial intelligence capabilities become more powerful and widespread, evolving and expanding the very nature of existing cybercrime threats while simultaneously spawning new threats. Indeed, there is reason to expect that intrusions enabled by the growing use of AI among cybercriminals will be finely targeted at the complex vulnerabilities created by AI systems and become more effective at exploiting the weaknesses left in their wake [15]. The emergence of machine learning algorithms, in particular, has effectively boosted adversary capabilities to run complex and repeatable problem-solving operations against unfortified positions without human intervention, providing cybercriminals with technical scalability and automation which has historically been beyond their reach. The ability of cybercriminals to more intelligently and systematically assault numerous targets at once will likely exacerbate an already challenging problem facing cyber security practitioners in which criminals must only find one flaw in a vast system, whereas database and systems administrators must account for all possible weaknesses to protect system integrity [25]. Even the most inept cyber-criminal need only exploit a single path of vulnerability among the complex and increasing number of data ingestion points, whereas data guardians face the increasingly difficult task of protecting against all conceivable threats to privacy [26].

2.2 Threat detection and attribution

While cybercrime offenses against privacy may in some ways be synonymous with traditional non-violent "street" crimes, such as those against property, because they involve the theft, corruption, or destruction of assets held and valued by a property owner, there may be a tendency to address them like ordinary crimes. However, the nature of technology-based privacy crimes varies in several important ways. Chief among these is the fact that cybercrimes often carry an inherently lower risk of detection, due to significant spatial separation and temporal distance between offenders and victims. Additionally, privacy-related offenses may also be obscured due to their velocity, automation, and complexity [27]. Thus, the adoption of new computing innovations and methods, such as machine learning, by cybercriminals will likely continue to challenge existing cybercrime detection and attribution methods. In particular, cyber-assaults against distributed systems may be of such increasing scale and complexity that forensic detection and attribution efforts will suffer markedly. Research has already shown cybercriminals to be savvy, having migrated away from easily detectable attacks that were recently commonplace toward more stealthy aggressions that are often indistinguishable [24].

For similar reasons, cybercrime threats will presumably expand and diversify as a natural byproduct of the automation computing innovations have permitted. In this regard, human capital costs of cybercriminals attempting intrusions into databases containing personal information are likely to decline as they leverage the scalable use of AI systems to complete tasks that would ordinarily require extensive human labor, intelligence and expertise. Those cost savings might naturally translate into expanding the pool of actors with which to initiate attacks, increasing the rate at which attacks are carried out, and growing the set of prospective targets. Thus, the acquisition of AI capabilities among cybercriminals will expand their operations to spawn new attacks that would be otherwise impractical for humans. Malicious actors will purposely target and exploit the growing multitude of vulnerabilities of AI systems deployed by those entrusted with stewardship and fortification of data, thereby deepening the threat to the privacy of individuals represented in such data.

3. Evolving privacy methods

While the influence and intrusion of technology into the public sphere has unintentionally created new opportunities for cyber victimization, various approaches to counter emerging threats have developed and evolved out of privacy requirements engineering. These methods have enabled the design, analysis, and integration of security and privacy requirements during systems implementation for traditional and cloud architectures to better support and protect data [28]. Further, novel privacy definitions have been created, resulting in several systematic approaches to minimize the likelihood of unintended data disclosures. Differential Privacy represents one of the newest, and perhaps most promising, privacy definitions aimed at preserving the privacy of individuals and groups whose data is published and/or accessible for public- and private-sector research and data analysis, as well as product and service development and enhancement. Yet a variety of other techniques continue to persist.

3.1 Prior anonymization techniques

As the scale of consumable data generated by society has grown, so too have the mechanisms for shielding the information and individuals represented in such data. Historically, curators of large databases attempted to protect individual privacy through the de-identification of datasets using a variety of algorithmic data anonymization techniques. These have included stripping or suppressing identifying information such as names, dates of birth, and other personal information out of data that is released for consumption, or through replacement of some data values with generalized quasi-identifiers. In effect, the data elements generated from these processes have represented approximations of data or a broad category of values to achieve the property of "k-anonymity"—anonymization resulting from data that is indistinguishable from that produced by another individual in the same dataset [29]. Through these practices, curators reasonably believed anonymity could be assured—that personal identifiable information (PII) contained within the data could not be distinguished or used to discover the identity of individuals or groups of individuals represented in the data [30]. However, we now know that these earlier methods for protecting individual privacy have been afflicted with vulnerabilities, resulting in "de-identified" datasets being prone to exploitation or attack, particularly where the value of sensitive attributes is not diverse enough or when sufficient background knowledge is known by would-be attackers [31]. In such circumstances, individuals might face unintentional risk of cybercrime victimization and identification resulting from inference attacks and algorithms deployed against databases to reconstruct case-specific identities through whatever limited, sensitive data is contained in a given database, or through the fusion of extracted data with external sources [32].

Numerous examples have been cited where de-identified data published for legitimate use was nevertheless systematically exploited to uncover individual identities (see [33–35]). Though some privacy breaches may not involve nefarious intent and therefore result in relatively benign consequences, the growing number of intentionally harmful and illegal privacy intrusions should elicit concern among privacy advocates and information security practitioners. Further, subsequent research has also revealed that not all k-anonymity algorithms provide uniform, privacy-preserving protections [36] and that some can inadvertently distort data to a point where both its integrity and utility are appreciably diminished [37]. Thus, it is clear that prior efforts to counter privacy risks have not gone far enough. While more recent techniques such as l-diversity and t-closeness have incrementally advanced the security of personal-level data, they may also be vulnerable to exploitation as the liquidity of data and proliferation of artificial intelligence in today's contemporary world continue to advance [38, 39]. Yet, despite these notable concerns, many of the deficient database de-identification techniques referenced above, which fail to truly anonymize participants and protect their confidentiality, continue to persist as commonplace practices in commercial industries and the larger research community [34].

3.2 Emergence of Differential Privacy

Recognizing the need for a more robust privacy approach, Differential Privacy was developed in the early 2000s. While it was not explicitly intended to guard against cybercrime, Differential Privacy represents a deliberate attempt to overcome many of the foreseeable privacy challenges identified above by seeking true anonymity in datasets. With this definition and the use of differentially private processes, personal information can, in theory, be more adequately protected from cybercrime activity by avoiding the availability or release of raw data and instead enabling a replica database upon which queries are run containing modified (but statistically similar) versions of person-level data. Thus, Differential Privacy represents an enhanced level of privacy protection in the evolving data security model, resulting in virtually no disclosure risk. It achieves this by obscuring individual identities with the addition of mathematical "noise" to particular data elements, consequently concealing a small sample of each individual's data [40]. According to its proponents, Differential Privacy virtually guarantees that the removal or addition of a single database item does not appreciably affect the outcome or validity of any analysis. Stated another way, this data perturbation technique ensures that the probability of a statistical query producing a given result is virtually the same whether it is conducted on an unadulterated dataset or one containing modified or synthetic data [40]. Thus, the true benefit to Differential Privacy is that there is quantifiably lower risk associated with its use over alternative methods aimed at systematically safeguarding personal data. In turn, individuals' data should be more rigorously defended from theft or illegitimate use when differentially private methods are used.

Because Differential Privacy was conceived as a more rigorous definition of anonymizing data and protecting confidentiality than prior methods, its popularity has grown in recent years, with several commercial entities enabling Differential Privacy algorithms for use on a massive scale for data generated in the private sector. For example, Apple has intentionally deployed Differential Privacy techniques to discover and analyze usage patterns of large numbers of iPhone users without compromising the privacy of individuals [40]. In this instance, Differential Privacy algorithms executed by Apple analyze iOS user data with the published goal of improving and enhancing end-user experiences with various iOS applications such as iMessage (text messaging), through which functions such as auto-correct, suggested words and phrases, and emojis can become more intuitive [41]. In a similar example of commercial use, Google has employed Differential Privacy algorithms in its analyses of Chrome web-browser usage to discover the prevalence of malicious software hijacking computer and application settings without user knowledge [42].

There has even been expanded use of Differential Privacy in the public sphere, with the U.S. Census Bureau recently announcing its plan to more rigorously protect the confidentiality of individual-level data than in years past. Prior to the most recent census, this federal agency attempted to obscure person-level information by substituting raw data beneath the census block level with comparable data from another block to ensure the validity of population-level statistics. However, beginning with the 2020 Census, "noise" will be purposely injected into all data emanating below the state geographic level [43] to achieve "advanced disclosure protections" [44]. This instance of Differential Privacy use represents one of the first by a federal agency broadly responsible for the collection and provision of data for public use, and is likely to serve as a possible model for other federal, state, and local data stewards.

Given its intent, generally positive reviews, and notable use in a handful of public and private sector instances, it is somewhat remarkable that Differential Privacy has failed to gain widespread adoption as a data protection measure since its introduction in 2006. Though Differential Privacy has indeed become an information security standard with database computation and analysis in computer science research, resulting in numerous algorithms aimed at strengthening privacy, practitioner adoption of Differential Privacy in applied settings has been slow to gain traction [45]. Similarly, while Differential Privacy has indeed spawned important new lines of data privacy research, much of that work has been theoretical or simulated and proven to be less suitable for application to real-world situations [4]. To date there have been few empirical examinations of the practical application of Differential Privacy, despite the existence of important concerns surrounding its viability, including possible tradeoffs that arise between achieving heightened privacy protections and preserving the utility of data produced through differentially private queries [46]. Despite the obvious and substantial lag between the emergence of Differential Privacy as a definition worthy of research and its acceptance as a pragmatic and commonly employed approach in real-world scenarios, it is important to consider the relevance and utility of Differential Privacy as a possible cybercrime countermeasure in anticipation that its use will someday become pervasive.

4. Cybercrime prevention through Differential Privacy

Though Differential Privacy is applicable to a number of industries and scenarios, its potential as a cybercrime prevention and risk mitigation measure is intriguing and warrants deeper exploration. From a criminological standpoint, differentially private approaches might best be deployed as technical situational crime prevention (SCP) measures to deter prospective attacks against sensitive data, or at the very least, minimize their harms. Generally speaking, situational crime prevention represents a data-driven approach to reduce the physical opportunities for crime by concentrating on the specific conditions, settings, and circumstances which produce the conditions favorable to criminality [47]. Further, the approach explicitly suggests that crime prevention can only be accomplished by systematically analyzing the details of a given crime problem and then introducing strategies for blocking, reducing or removing the opportunities that enable a particular crime to take place [14]. Thus, the most viable strategy to combat crime is through the informed management, design, and manipulation of a particular environment that would ordinarily be conducive to crime [48]. While SCP has mostly been utilized to examine and respond to traditional forms of criminality, such as burglary, robbery, and theft, it has direct applicability to cybercrime, given that acts of cybercrime share many similarities with property crimes. By examining important contextual attributes associated with specific cybercrime events, such as the technical means and steps through which an attack on data may be committed and how a database containing private information may be made less attractive or be better protected, cybersecurity practitioners can develop competent proactive strategies to reduce the presence and attractiveness of criminal possibilities for would-be offenders [14].

Situational Crime Prevention efforts are generally intended to achieve three goals: increase the overall risk to criminals, increase the effort they would be required to expend to engage in a crime, and decrease the reward associated with an act of crime [49]. In practice, exploration of a given network or computerized system through the perspective of situational crime prevention might first enable the identification of various targets that represent higher-value for cybercriminals. In turn, those high-value targets would be the first and most likely to receive heightened privacy protections. For example, databases that contain sensitive information about individuals or groups which, if disclosed, might hold potential monetary value and likely result in physical or financial harm, would be ideal candidates for Differential Privacy protections. Once identified, possible cybercrime targets might be "hardened" and made less attractive through the intentional adulteration of data in an effort to obscure personal information. The intent of this tactic would be to reduce the likelihood of an attack, because the risk and effort for a cybercriminal initiating an assault on that target would be considerably greater than in situations where differentially private techniques are not used.

The act of safeguarding data clearly carries direct costs for data stewards and information security practitioners, but attacks against data also carry similar costs for the attacker, both in terms of the resources required to mount an attack and potential costs if an attack is detected and subsequently punished. Unless the expected return from an attack is greater than the risk-adjusted costs of the attack, the attack will be uneconomical and become a less attractive target for an offender. Thus, the injection of noise into an otherwise high-value, sensitive dataset through

Differential Privacy algorithms might ensure that attackers would have to work harder and still be unable to derive much, if any, value from stolen data, despite the data still remaining useful for legitimate purposes. As a data perturbation method, Differential Privacy stands to more securely protect the privacy of individuals and appreciably diminish the utility of the entire corpus of stolen data, thus negating an attacker's reward motive. With advance knowledge of the use of differentially private techniques on high-value databases, cybercriminals might altogether be deterred from exerting the effort to wage an attack, given the minimal value of the data relative to the cost of waging such an attack.

5. Practical challenges

Despite confidence in Differential Privacy as a promising new tool in the war against cybercrime, it is not a panacea. A number of practical concerns remain that may slow the adoption of this approach in the near-term and challenge its use as a viable cybercrime countermeasure. Each of the following challenges should be examined more thoroughly to guide future decision-making for the use of Differential Privacy in real-world settings. Chief among these concerns are the trade-offs that accompany the use of Differential Privacy, specifically, where the costs associated with using differentially private methods are balanced against the benefits of doing so. Second, while the likelihood of privacy intrusions originating external to a given system might fall with the use of Differential Privacy, there is a possible shift in risk from external to internal threats that is likely to accompany the use of Differential Privacy in a variety of applied settings. Similarly, as use of Differential Privacy grows, adversaries will also be increasingly more likely to take advantage of advances in computing power, launching a virtual "arms race" between cybercriminals and those responsible for protecting sensitive data. Lastly, but perhaps most limiting for the use of Differential Privacy, particularly in crime and justice settings, there remains a very real concern about the practical challenge of resourcing the skilled human capital needed to develop, enable, and continually support Differential Privacy techniques.

5.1 Tradeoffs

An important implication of Differential Privacy is that its use results in two significant tradeoffs that should be factored into decisions regarding whether, when, and how to use the method. In the first tradeoff, the validity or accuracy of a given set of data may be reduced with a corresponding attempt to increase privacy. For example, the near-guarantee of total anonymity in a dataset can only be attained at some proportional reduction to the utility of that dataset. This challenge is commonly referred to as the "privacy budget" [50]. In this regard, tipping the scales in favor of greater privacy protections by injecting noise into data will provide a clear privacy benefit to the individuals whose personal information is contained in a given database. However, the adulteration of data resulting from a differentially private technique may also unintentionally produce imprecise statistical measures of a given phenomenon and lead to invalid conclusions derived from analysis of the data. The risk associated with this situation is that conclusions drawn from adulterated data under legitimate use scenarios, either by researchers or practitioners, might be faulty, because they are based on inaccurate data.

One cautionary example of this challenge is a pharmacogenetic study conducted by Fredrikson et al. [50]. The research evaluated the clinical effectiveness of a commonly prescribed blood-thinner using machine-learning models, while differentially private algorithms were enabled to significantly reduce privacy risk for study participants. While the study yielded success in appreciably reducing privacy risk for study participants, according to the data, that success came at an increased risk of patient adverse health events and mortality. Though the study itself was simulated to examine the impact of Differential Privacy on a real-world clinical situation, the possible implications are clear; using differentially private algorithms to produce synthetic data may lessen privacy risks, but consequently result in a variety of unintended consequences to the conclusions of research, or in a worst-case scenario, to the same people Differential Privacy is meant to protect.

In addition to the tradeoff concern relating to the privacy budget, Differential Privacy also requires a tradeoff between the costs of deploying the privacy protections and the relative value of the data assets being protected. The values of data assets differ widely. Some targets might contain high-value, sensitive information, such as personal identifiers, credit card information, passwords, social security numbers, and insurance information that can be used maliciously to steal an identity or file false Medicare claims. Cybercriminals would likely view these targets as attractive and initiate attacks against the databases to steal such information. Therefore, databases containing highly sensitive data need extremely high-assurance protections. Other targets may contain personal data but of a less sensitive variety, including Netflix subscriptions, personal shopping preferences, search term use, or website visits. The value of these data may have lower transactional value for cybercriminals looking to exploit personal information. Thus, datasets containing these sources of information would presumably require weaker assurance protections.

A scenario where both high-and low-value assets are guarded requires that hazard-based decisions be made about the effort devoted to protecting each set of assets from cybercriminals. For example, security practitioners should explore what must be done to sufficiently protect high-assurance assets from possible intrusion, and what minimum level of effort would be required to protect low-assurance assets. Treating low-assurance assets the same as high-value would lead to the irrational use of resources. Therefore, practitioners should carefully consider tradeoffs to the privacy budget and efforts required to protect assets when choosing to implement differentially private approaches.

5.2 Shifting risk and the impending arms race

While the adoption of Differential Privacy techniques may provably strengthen defenses against traditional cybercrime threats directed at the theft of personal information from a database, their use may also coincide with a sizeable shift in where risks originate and how they evolve. For instance, there is already mounting concern among researchers and practitioners that new innovations and technology advances will transform the very nature of systems integrity and vulnerability, particularly with the growth of artificial intelligence, which will result in a "double-barreled threat" to high-value data repositories [14, 51]. In the traditional cybercrime model, criminal threats are generally thought to arise from an external source, spatially distant from the data being protected. However, internal threats to systems and data are now garnering additional attention, as cybercrime attacks are being more frequently initiated by organizational insiders [52]. The growing likelihood and simultaneous nature of these dual threats significantly increases the effort necessary to keep an infrastructure and its data secure, which will represent a significant ongoing challenge for many industries and organizations already struggling to provide robust information security [51].

Further, as Differential Privacy continues its incremental expansion beyond the realm of research toward use in applied settings, the resources and costs required for enabling Differential Privacy and other sophisticated privacy protections will also evolve. So too will the costs for cybercriminals intent on defeating the stronger protections afforded by differentially private systems. Cybercriminals are already taking advantage of more powerful computational resources and sophisticated approaches, requiring the investment of data guardians to continue increasing proportionally to keep pace. As a result of the commodification of computing technology, there is a brewing cybercrime "arms race" where information security practitioners will be constantly expected to respond in tit-for-tat fashion to complex and powerful threats from hostile actors [53]. As a result, to avoid having information security devolve into a never-ending game of "whack-a-mole" to combat emerging threats, individuals responsible for data security policy and practice must develop comprehensive strategies for data management and the use of privacy-preserving tools like Differential Privacy. However, the creation of such policies requires careful consideration of the origin and nature of threats to the data for which organizations have responsibility.

5.3 Resource constraints

Finally, and despite its potential as an automated method of systematically safeguarding data, Differential Privacy, much like artificial intelligence (see [54]), will only be as useful as the skilled humans that enable and support it. Unfortunately, some of the most pressing information security concerns facing a majority of organizations today include the limited number of skilled security personnel employed and the number who are readily available for employment [54]. While Differential Privacy strategies offer the realistic promise of protecting data for organizations that cater to consumers, significant barriers to the implementation and use of advanced privacy-enhancing technologies remain for organizations and agencies in the public sector that curate data for the most vulnerable populations, such as patients, prisoners, the disabled, and juveniles. Differential Privacy use to date has taken place primarily in the private sector, within organizations that have the financial and intellectual resources to pursue novel and costly privacy protections. However, research suggests that federal agencies do not have the relevant expertise or resourcing to implement differential privacy for the data they curate [55]. This is evident in the fact that to this point the U.S. Census Bureau is the only federal agency known to have initiated a systematic effort to employ Differential Privacy with the data it curates. The increasing sophistication of prospective cybercriminals and growing complexity of privacy enhancing technologies, including Differential Privacy algorithms needed to protect sensitive data, requires a level of data security expertise and sophistication that is simply not readily available throughout the federal public sector. In turn, this limitation is likely to be amplified at the state and local agency level, where funding for and expertise in skilled information security personnel are even more severely restricted than with the federal government.

Though expertise and a skilled labor force will become more common with the pervasiveness of Differential Privacy and other privacy-preserving technologies, it is sure to take time. And even then, organizations in the public sector may continue to face the difficulty of competing against private sector organizations to hire and retain personal capable of developing and enabling the use of robust privacy measures on vulnerable data.

6. Conclusion

This unprecedented era of technology "connectedness" and "big data" has virtually assured that we will never be left entirely alone, and that our views of privacy will be forever changed by the digital means through which we now interact with the world around us [56]. Similarly, this new way of life raises the ever-present specter of devastating privacy risks resulting from cybercrime that compromises or steals the personal data we all generate and share. Given its potential as a pragmatic tool for organizations and data stewards in the war against growing cybercrime threats, Differential Privacy fits well within a situational crime prevention framework and possibly represents a model to guide future privacy requirements engineering and protections directed exclusively at the issue of cybercrime. Therefore, policymakers and practitioners would be well-served to engage in empirical exploration of the implications that Differential Privacy and situational crime prevention collectively have on existing and new forms of cybercrime that are likely to emerge in the future.

Notwithstanding the practical challenges identified above, there is a continuing need for exploration and development of data privacy and disclosure methods that match our shifting data culture and also maintain the public's trust in institutions and industries [4]. The pursuit of these aims can be achieved through greater consideration of securing software systems early in development and implementation lifecycles and through a more dedicated focus on expanding privacy requirements engineering research [28, 57]. Additional attention should be directed more specifically at Differential Privacy as a unique design and implementation method. Research on the practical application and limitations of privacy enhancing technologies like Differential Privacy within the context of cybercrime remains necessary. In this regard, future studies might wish to employ a "no-free-lunch theorem" and investigate some of the popular misconceptions about Differential Privacy and its vulnerabilities, such as making no assumptions about how data are generated, that it protects personal information despite an attacker having knowledge of other individuals represented in the data, and that it is defensible to arbitrary background knowledge [58]. Doing so would ensure that subsequent use of Differential Privacy does not inadvertently contribute to future privacy-related challenges.

Generally speaking, most research exploring risks to individual privacy have been aimed squarely at consumer protection in the private sector. And while the average consumer should be cautious about the risks associated with sharing data for commercial use, there are other groups for which data privacy becomes a more considerable challenge. Vulnerable populations such as patients, children, the indigent, the elderly, inmates, undocumented immigrants, the civilly committed, and the mentally ill, are some of the most frequently studied populations, but are among the least likely to have the sufficient protections from data privacy intrusions. Efforts should be made to correct this imbalance by finding opportunities to make costly privacy-enhancing technologies available to public sector agencies.

There is also a significant need and opportunity for cross-disciplinary collaboration with respect to cybercrime and privacy-related research. Scholars from technical and social science disciplines are encouraged to join forces to expand the scope and breadth of research on the many threats to privacy which stem from cybercrime. They should also work together to investigate the variety of promising opportunities for preventing and responding to cybercrime threats, including Differential Privacy. Doing so would undoubtedly contribute to the development and spread of more appropriate and accessible approaches to the preservation of privacy.

Ultimately, before moving forward with any Differential Privacy or any other privacy-enhancing technologies, data scientists, researchers, and practitioners should collaborate and carefully explore the consequences of this evolution in data protection. Additional resources and effort should be dedicated to the careful appraisal of privacy protections for person-level data in a variety of public and private scenarios. Failure to do so will likely result in more frequent and severe cybercrime breaches of critical infrastructure and significant privacy implications for individuals and groups whose data is widely available and easily accessible.

Author details

William Stadler
Saint Martin's University, Lacey, WA, USA

*Address all correspondence to: wstadler@stmartin.edu

References

[1] Mivule K. Utilizing noise addition for data privacy, an overview. In: Proceedings of the International Conference on Information and Knowledge Engineering (IKE 2012). 2012. pp. 65-71

[2] Brundage M, Avin S, Clark J, Toner H, Eckersley P, Garfinkel B, et al. The malicious use of artificial intelligence: Forecasting, prevention, and mitigation. Design Direction [Internet]. 2018:1-101. Available from: https://arxiv.org/pdf/1802.07228.pdf [Accessed: 24 March 2020]

[3] Diamond B, Bachmann M. Out of the beta phase: Obstacles, challenges, and promising paths in the study of cyber criminology. International Journal of Cyber Criminology [Internet]. 2015;**9**(1):24-34. Available from: http://www.cybercrimejournal.com [Accessed: 03 April 2020]

[4] Snoke J, Bowen CM. Differential privacy: What is it? AMSTAT News [Internet]. 2019. Available from: https://magazine.amstat.org/blog/2019/03/01/differentialprivacy/ [Accessed: 03 April 2020]

[5] Wilner AS. Cybersecurity and its discontents: Artificial intelligence, the Internet of Things, and digital misinformation. International Journal: Canada's Journal of Global Policy Analysis [Internet]. 2018;**73**(2):308-316. DOI: 10.1177/0020702018782496. Available from: http://journals.sagepub.com [Accessed: 17 April 2020]

[6] Dwork C. Differential privacy. In: Lecture Notes in Computer Science (Including Subseries Lecture Notes in Artificial Intelligence and Lecture Notes in Bioinformatics). Berlin, Heidelberg: Springer; 2006. pp. 1-12

[7] Dwork C. Differential privacy: A survey of results. In: Agrawal M, Du D, Duan Z, Li A, editors. Theory and Applications of Models of Computation. Berlin, Heidelberg: Springer; 2008. pp. 1-19

[8] Clarke RV. Technology, criminology and crime science. Crime and Deviance in Cyberspace. 2004;**10**(1):441-450

[9] Holt TJ, Bossler AM. Cybercrime in Progress: Theory and Prevention of Technology-Enabled Offenses. New York: Routledge; 2015

[10] Gordon S, Ford R. On the definition and classification of cybercrime. Journal in Computer Virology. 2006;**2**(1):13-20

[11] Anderson R, Barton C, Böhme R, Clayton R, van Eeten MJG, Levi M, et al. Measuring the cost of cybercrime. In: Bohme R, editor. The Economics of Information Security and Privacy. Berlin Heidelberg: Springer; 2013. pp. 265-300

[12] Tabansky L. Cybercrime: A national security issue? Military and Strategic Affairs. 2012;**4**(3):117-136

[13] Subramanian R, Sedita S. Are cybercrime laws keeping up with the triple convergence of information, innovation and technology? Communication IIMA [Internet]. 2006;**6**(1):39-50. Available from: https://scholarworks.lib.csusb.edu/ciima; https://scholarworks.lib.csusb.edu/ciima/vol6/iss1/4 [Accessed: 15 April 2020]

[14] Hinduja S, Kooi B. Curtailing cyber and information security vulnerabilities through situational crime prevention. Security Journal [Internet]. 2013;**26**(4):383-402. Available from: www.palgrave-journals.com/sj/ [Accessed: 16 April 2020]

[15] Khisamova ZI, Begishev IR, Sidorenko EL. Artificial intelligence and problems of ensuring cyber

security. International Journal of Cyber Criminology. 2019;**13**(2):564-577

[16] Finkle J, Chatterjee S, Maan L. EBay asks 145 million users to change passwords after cyber attack. Reuters [Internet]. 2014. Available from: https://www.reuters.com/article/us-ebay-password/ebay-asks-145-million-users-to-change-passwords-after-cyber-attack-idUSBREA4K0B420140521 [Accessed: 15 April 2020]

[17] Silver-Greenberg J, Goldstein M, Perlroth N. JPMorgan Chase hacking affects 76 million households. The New York Times [Internet]. 2014. Available from: https://dealbook.nytimes.com/2014/10/02/jpmorgan-discovers-further-cyber-security-issues/ [Accessed: 15 April 2020]

[18] Cieply M, Barnes B. Sony cyberattack, first a nuisance, swiftly grew into a firestorm. The New York Times [Internet]. 2014. Available from: https://www.nytimes.com/2014/12/31/business/media/sony-attack-first-a-nuisance-swiftly-grew-into-a-firestorm-.html [Accessed: 15 April 2020]

[19] Swinhoe D. The 14 biggest data breaches of the 21st century. CSO Online [Internet]. 2020. Available from: https://www.csoonline.com/article/2130877/the-biggest-data-breaches-of-the-21st-century.html [Accessed: 15 April 2020]

[20] Perlroth N. D.N.C. says it was targeted again by Russian hackers after '18 election. The New York Times [Internet]. 2019. Available from: https://www.nytimes.com/2019/01/18/technology/dnc-russian-hacking.html [Accessed: 16 April 2020]

[21] Mazzei P. F.B.I. to Florida lawmakers: You were hacked by Russians, but don't tell voters. The New York Times [Internet]. 2019. Available from: https://www.nytimes.com/2019/05/16/us/florida-election-hacking-russians-fbi.html [Accessed: 15 April 2020]

[22] Broadhurst R, Grabosky P, Alazab M, Bouhours B, Chon SK. Crime in cyberspace: Offenders and the role of organized crime groups. SSRN Electronic Journal. 2013:1-42. Available from: https://papers.ssrn.com/sol3/papers.cfm?abstract_id=2211842 [Accessed: 24 March 2020]

[23] Cline S, Aronoff J. With great power comes great responsibility: Utilizing privacy technology for the greater bad. arXiv:200100226 [Internet]. 2019;**1**. Available from: http://arxiv.org/abs/2001.00226 [Accessed: 26 March 2020]

[24] Provos N, Rajab MA, Mavrommatis P. Cybercrime 2.0: When the cloud turns dark. Communications of the ACM. 2009;**52**(4):42-47

[25] Dickson B. The darker side of machine learning. TechCrunch [Internet]. 2016. Available from: https://techcrunch.com/2016/10/26/the-darker-side-of-machine-learning/ [Accessed: 06 April 2020]

[26] Herley C. Security, cybercrime, and scale. Communications of the ACM. 2014;**57**(9):64-71

[27] Lallement P. The cybercrime process: An overview of scientific challenges and methods. International Journal of Advanced Computer Science and Applications. 2013;**4**(12):72-78

[28] Pattakou A, Kalloniatis C, Gritzalis S. Security and privacy requirements engineering methods for traditional and cloud-based systems: A review. In: Dini P, editor. Proceedings of the CLOUD COMPUTING 2017 8th International Conference on Cloud Computing, GRIDs, and Virtualization. Athens, Greece; 2017

[29] Sweeney L. k-anonymity: A model for k-anonymity: A model for protecting privacy. International Journal of Uncertainty, Fuzziness and Knowlege-Based Systems. 2002;**10**(5):557-570

[30] Sweeney L. Weaving technology and policy together to maintain confidentiality. Journal of Law, Medicine & Ethics [Internet]. 1997;**25**(2-3):98-110. DOI: 10.1111/j.1748-720X.1997.tb01885.x. Available from: http://journals.sagepub.com [Accessed: 26 March 2020]

[31] Machanavajjhala A, Kifer D, Gehrke J, Venkitasubramaniam M. ℓ-diversity: Privacy beyond k-anonymity. ACM Transactions on Knowledge Discovery from Data [Internet]. 2007;**1**(1):3. Available from: http://portal.acm.org/citation.cfm?doid=1217299.1217302 [Accessed: 24 March 2020]

[32] Ciriani V, Capitani di Vimercati S, Foresti S, Samarati P. In: Yu T, Jajodia S, editors. Secure Data Management in Decentralized Systems. New York, NY: Springer; 2007. pp. 291-321

[33] Barbaro M, Zeller T. A face is exposed for AOL searcher no. 4417749. The New York Times [Internet]. 2006. Available from: http://www.nytimes.com/2006/08/09/technology/09aol.html?ei=5090&e [Accessed: 26 March 2020]

[34] Heffetz O, Ligett K. Privacy and data-based research. The Journal of Economic Perspectives. 2014;**28**(2):75-98

[35] Narayanan A, Shmatikov V. Robust de-anonymization of large datasets (how to break anonymity of the Netflix Prize dataset). arXiv:cs/0610105 [Internet]. 2006:1-24. Available from: http://arxiv.org/abs/cs/0610105 [Accessed: 26 March 2020]

[36] Ayala-Rivera V, McDonagh P, Cerqueus T, Murphy L. A systematic comparison and evaluation of k-anonymization algorithms for practitioners. Transactions on Data Privacy. 2014;7:337-370

[37] El Emam K, Dankar FK. Protecting privacy using k-anonymity. Journal of the American Medical Informatics Association. 2008;**15**(5):627-637

[38] Li N, Li T, Venkatasubramanian S. T-closeness: Privacy beyond k-anonymity and-diversity. In: IEEE International Conference on Data Engineering. 2007

[39] Wong RCW, Fu AWC, Wang K, Yu PS, Pei J. Can the utility of anonymized data be used for privacy breaches? ACM Transactions on Knowledge Discovery from Data [Internet]. 2011;5(3):1-24. doi 10.1145/1993077.1993080. Available from: https://dl.acm.org [Accessed: 15 April 2020]

[40] Green M. What is Differential Privacy?—A Few Thoughts on Cryptographic Engineering [Internet]. 2016. Available from: https://blog.cryptographyengineering.com/2016/06/15/what-is-differential-privacy/ [Accessed: 24 March 2020]

[41] Apple Inc. Differential Privacy Team. Learning with Privacy at Scale [Internet]. 2017. Available from: https://machinelearning.apple.com/2017/12/06/learning-with-privacy-at-scale.html [Accessed: 27 March 2020]

[42] Erlingsson Ú. Learning statistics with privacy, aided by the flip of a coin. Google Online Security Blog [Internet]. 2014. Available from: https://security.googleblog.com/2014/10/learning-statistics-with-privacy-aided.html [Accessed: 25 March 2020]

[43] National Conference of State Legislatures. Differential Privacy for Census Data Explained [Internet]. 2020 Available from: https://www.ncsl.org/research/redistricting/differential-privacy-for-census-data-explained.aspx [Accessed: 26 March 2020]

[44] United States Census Bureau. Disclosure Avoidance and the 2020 Census [Internet]. 2020. Available from: https://www.census.gov/about/policies/privacy/statistical_safeguards/disclosure-avoidance-2020-census.html [Accessed: 30 March 2020]

[45] Machanavajjhala A, He X, Hay M. Differential privacy in the wild: A tutorial on current practices & open challenges. Proceedings of the VLDB Endowment. 2016;**9**(13):1611-1614

[46] Hsu J, Gaboardi M, Haeberlen A, Khanna S, Narayan A, Pierce BC, et al. Differential privacy: An economic method for choosing epsilon. In: Proceedings of the Computer Security Foundations Symposium [Internet]. IEEE Computer Society; 2014. pp. 398-410. Available from: http://arxiv.org/abs/1402.3329 [Accessed: 25 March 2020]

[47] RVG C. Situational crime prevention: Theory and practice. British Journal of Criminology. 1980;**20**(2):136-147. Available from: https://heinonline.org/HOL/Page?handle=hein.journals/bjcrim20&id=148&div=19&collection=journals [Accessed: 16 April 2020]

[48] Clarke RV. Situational Crime Prevention: Successful Case Studies. 2nd ed. Guilderland, New York: Harrow and Heston; 1997

[49] Clarke RV, Homel R. A revised classification of situational crime prevention techniques. In: Lab SP, editor. Crime Prevention at a Crossroads. Cincinnati, OH: Anderson; 1997

[50] Fredrikson M, Lantz E, Jha S, Lin S, Page D, Ristenpart T. Privacy in pharmacogenetics: An end-to-end case study of personalized warfarin dosing. In: Proceedings of the 23rd USENIX Security Symposium [Internet]. San Diego, CA; 2014. p. 17. Available from: https://www.usenix.org/conference/usenixsecurity14/technical-sessions/presentation/fredrikson_matthew [Accessed: 27 March 2020]

[51] Higgins GE. Cybercrime: An Introduction to an Emerging Phenomenon. New York: McGraw-Hill; 2009

[52] Greitzer FL, Moore AP, Cappelli DM, Andrews DH, Carroll LA, Hull TD. Combating the insider cyber threat. IEEE Security and Privacy. 2008;**6**(1):61-64

[53] Mansfield-Devine S. The malware arms race. Computer Fraud & Security. 2018;**2**:15-20

[54] Maher D. Can artificial intelligence help in the war on cybercrime? Computer Fraud & Security. 2017;**2017**(8):7-9

[55] Reiter JP. Differential privacy and federal data releases. Annual Review of Statistics and Its Application [Internet]. 2019;**6**:85-102. Available from: https://www.annualreviews.org/doi/abs/10.1146/annurev-statistics-030718-105142 [Accessed: 24 March 2020]

[56] Jerome J. Big data: Catalyst for a privacy conversation. Indiana Law Review [Internet]. 2014;**48**(1):213-242. Available from: https://heinonline.org/HOL/Page?handle=hein.journals/indilr48&id=229&div=12&collection=journals [Acessed: 16 April 2020]

[57] Mouratidis H, Argyropoulos N, Shei S. Security requirements

engineering for cloud computing: The secure tropos approach. In: Domain-Specific Conceptual Modeling: Concepts, Methods and Tools. Switzerland: Springer International Publishing; 2016. pp. 357-380

[58] Nguyen HH, Kim J, Kim Y. Differential privacy in practice. Journal of Computing Science and Engineering. 2013;7(3):177-186

Chapter 8

A Review of Several Privacy Violation Measures for Large Networks under Active Attacks

Tanima Chatterjee, Nasim Mobasheri and Bhaskar DasGupta

Abstract

It is by now a standard practice to use the concepts and terminologies of network science to analyze social networks of interconnections between people such as Facebook, Twitter and LinkedIn. The powers and implications of such social network analysis are indeed indisputable; for example, such analysis may uncover previously unknown knowledge on community-based involvements, media usages and individual engagements. However, all these benefits are not necessarily cost-free since a malicious individual could compromise privacy of users of these social networks for harmful purposes that may result in the disclosure of sensitive data that may be linked to its users. A natural way to avoid this consists of an "anonymization process" of the relevant social network. However, since such anonymization processes may not always succeed, an important research goal is to quantify and measure how much privacy a given social network can achieve. Toward this goal, some recent research works have aimed at evaluating the resistance of a social network against active privacy-violating attacks by introducing and studying a new and meaningful privacy measure for social networks. In this chapter, we review both theoretical and empirical aspects of such privacy violation measures of large networks under active attacks.

Keywords: social networks, privacy measure, active attacks, (k, ℓ)-anonymity, algorithmic complexity

1. Introduction

In recent years, social networks have become an indisputable part of people's lives. The emergence of such networks has altered how we interact with the world. A given individual's day-to-day activities like media consumption, job hunting and social interaction have changed, along with how businesses and other beneficial entities interact with them through marketing, advertising, and information diffusion. This has led to an unstoppable race of collecting information and interaction from social networks by researchers, governments, and business entities for various purposes. From a research point of view, social networks and their interaction mechanisms provide valuable insight in many fields of study, such as sociology, psychology, advertising, and recommendation systems. It is only natural that the information contained in these networks and the value they hold have been and will be targeted by bad actors for malicious activities. The importance of these networks and the value of information that can be retrieved from them have led social network researchers to take a closer look at methods to combat such bad actors as well as formulate network measures that can provide an insight to the privacy of these networks. In this survey, we will look at one such measure known as (k, ℓ)-anonymity [1] and will discuss some theoretical and empirical results regarding this measure.

1.1 Overview of the paper

Given the irrefutable importance of social networks in our daily lives and the ever increasing risk of compromising valuable personal data through privacy attacks against these networks, it is preferable to know how secure a given social network is against privacy attacks. This necessitates a deeper look into the types of privacy attacks and how to cope with them. There is an extensive literature on privacy preserving computational models in variety of application areas such as multi-party communications or distributed computing settings [2–6]. In this chapter, we focus on a specific type of attack known as *background-based active attack* and one measure that reflects the resistance of any given network against such attacks. The organization of the rest of the paper is as follows:

- In Section 2 we briefly discuss the notion of privacy in social networks and review some literature on privacy violating attacks on social networks. We also introduce the (k, ℓ)-anonymity privacy measure and some corresponding network measurement which are the basis for this measure.

- In Section 3 we review some basic terminologies and notations that will be used in formulation of the three problems introduced in Section 4.

- Section 4 contains three problems that arise from theoretical investigation of the (k, ℓ)-anonymity.

- Section 5 contains the results of an empirical study on the resistance of real-world social networks.

- Finally, we end this chapter with some concluding remarks in Section 6.

2. Privacy measures in social networks

We begin by discussing the mathematical structure that fit the most to represent social networks. A social network is often portrayed as a graph [7, 8] $G = (V, E)$ where V is a set of nodes representing the social members, and E is the set of edges portraying the relationship among these members. Both nodes and edges may have extra attributes, such as weights, that provide extra information about the nature of these social bonds (*e.g.*, trust or popularity); however, throughout this survey we will consider the simplest form of graphs, namely undirected and unweighted graphs, to model our social networks.

As we discussed in the previous section, the information that the social networks provide are invaluable. Due to the very nature of many social network applications, the identity of the members or the nature of relationship between members is quite sensitive and valuable. Thus, when releasing a social network we want to remove any attributes that may help identify these kinds of sensitive data. Assuming all members and their relationships are of high sensitivity, preventing *identity disclosure* or *link disclosure* becomes an important task. One popular method to prevent such disclosures is *anonymization*. In an anonymization process, we publish the network without identifying the corresponding nodes or potentially identifiable attributes. Even after anonymizing the network, we will still be releasing many informative attributes encoded by the network structure; for example, attributes such as node degree, connectivity, or other similar graph properties can still help the adversaries in compromising the user privacies of a published network.

Adversaries usually rely on background knowledge to compromise the privacy of published anonymized social networks. For understanding the failure of current privacy preservation methods such as anonymization, we need to have a proper model for the adversary background knowledge. Although it's challenging to have a comprehensive model of all possible types of adversary background knowledge, it is very useful to model the background knowledge via structural properties of networks such as node degrees, embedded subgraphs, node neighbors, etc. [9]. Backstrom et al. [10] were the first to introduce a category of attacks on anonymized social graphs. The models introduced in [10] are background-based attacks and are *widely* used in privacy analysis of social networks. The two main types of attacks are as follows.

1. *Passive attacks* in which the adversary will *not* modify the network by injecting new nodes, but instead will use the structural knowledge to detect the location of a *known* node. In this type of attacks, the adversary can benefit from the fact that most nodes in real social networks often belong to a small uniquely identifiable subgraph [10]. An adversary can then build a coalition with members of such subgraphs and attempt to re-identify the subgraphs in the anonymized published network, thus compromising the privacy of neighboring nodes.

2. *Active attacks* in which the adversary will choose an arbitrary set of target users, create new nodes and insert them into a social network in a way that they are connected to the target set and they form a distinguishable subgraph. After the anonymized version of the social network is published, the adversary can then use the subgraph as a *fingerprint* to re-identify the targeted users and compromise their privacy.

The authors in [10] also showed that it *is* possible to compromise the privacy of any social network of n nodes with high probability using *only* $O(\sqrt{\log n})$ attacker nodes. In a *passive attack*, adversary's structural knowledge will give her/him a global view of the network depending on the global structure of the network. It could pose a high privacy risk if an adversary were to combine this global view with the local structural knowledge obtained using an active attack. As an example, consider the network in **Figure 1**. If we only have global structural knowledge, it is not possible to differentiate the nodes v_3 and v_4 (*e.g.* , same node degrees, *etc.*). However, controlling just one extra node in the graph, such as the node v_1, provides local structural knowledge such as distances between nodes, and using the knowledge of the distance of v_1 from v_3 and v_4 ($d_{v_1,v_3} = 1$ and $d_{v_1,v_4} = 2$) one can easily differentiate node v_3 from node v_4.

There are several well-studied strategies for coping with active attacks on a social network [9, 11, 12] via addressing the anonymization process of the social network. However, in this chapter we will focus on a measure that evaluates how resistant a social network is against this type of privacy attack. Introduced by Trujillo-Rasua et al. [1], (k, ℓ)-anonymity is a novel and, to the best of our knowledge, the *only* privacy measure examining the structural resistance of a given graph against active attacks. The (k, ℓ)-anonymity is a measure based on metric representation of nodes, where k is a privacy threshold and l is the maximum number of

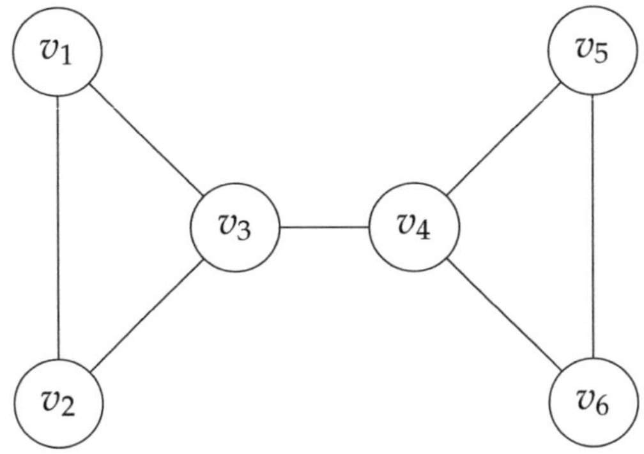

Figure 1.
A simple graph G *used in Section 2 to illustrate the high risk posed by combining knowledge gained by active and passive attacks.*

attacker nodes that may be inserted in the network. It was shown in [1] that graphs satisfying (k, ℓ)-anonymity can successfully deter adversaries controlling at most l nodes in the graph from re-identifying nodes with probability higher than $\frac{1}{k}$.

2.1 (k, ℓ)-anonymity

The (k, ℓ)-anonymity measure is based on a concept known as k-metric antidimension of graphs. To facilitate further discussions about the measure, we first introduce some notations and terminologies. For a simple connected graph $G = (V, E)$, where V is set of nodes and E is set of edges, let $dist_{v_i,v_j}$ denote distance (*i.e.* , number of edges in a shortest path) between the nodes v_i and v_j. Given and ordered set of nodes $S = \{v_1, \ldots, v_t\}$ and a node u we define the metric representation of u with respect to S as a vector $\mathbf{d}_{u,-S} = (dist_{u,v_1}, \ldots, dist_{u,v_t})$. Metric representations of nodes are closely related to the concept of a *resolving set* of a graph. Inspired by the problem of identifying an intruder in a network and introduced separately by Slater [13] and by Harary and Melter [14], a resolving set of graph provides recognition of every pair of nodes in graph.

Definition 1 (resolving set). Given a graph $G = (V, E)$, a subset $S \subseteq V$ is called a resolving set for G if, for each pair of nodes $(u, v) \in G$, there exist a node $x \in S$ such that $dist_{x,u} \neq dist_{x,v}$. A smallest-cardinality resolving set is called the metric basis, and its cardinality is referred to as the metric dimension of G.

The concepts of metric representation and resolving set inspired the introduction of another network measure known as *k-antiresolving set* that will be used as the founding base for (k, ℓ)-anonymity.

Definition 2 (k-antiresolving set). Given a graph $G = (V, E)$, $S \subset V$ is called a k-antiresolving set of G if k is the largest integer such that, for every node $v \in V \backslash S$, there exist at least $k - 1$ nodes $u_1, u_2, \ldots, u_{k-1} \in V \backslash S$ with the same metric representation with respect to S as v.

A k-antiresolving set of *minimum* cardinality is called a *k-antiresolving basis*, and its cardinality denotes the *k-metric antidimension* $adim_k(G)$ of G. Note that the k-antiresolving set may not exist for every k in a graph.

The (k,l)-anonymity measure is built upon the k-antiresolving set concept. Assume the adversary has gained control of a subset S of nodes in the graph G, where S is a k-antiresolving set for G. Then the adversary *cannot* uniquely re-identify any node based on the background knowledge (namely, the knowledge of metric representation of a node v with respect to S) with probability higher than $\frac{1}{k}$. (k, ℓ)-anonymity is formally defined as [1].

Definition 3 ((k, ℓ)-anonymity). A graph G under active attack satisfies (k, ℓ)-anonymity if k is the smallest positive integer so that the k-metric antidimension of G is less than or equal to l.

In the above definition, k is a parameter depicting the privacy threshold and l represents the maximum number of attacker nodes. It is safe to assume that number of attacker nodes l is significantly smaller than number of nodes present in the network as injecting attacker nodes or gaining control of existing nodes is difficult without being detected [15].

3. Basic terminologies and notations

For the exposition in the remainder of this chapter, we will need some notations and terminologies which we introduce here. Consider the (undirected unweighted) graph G in **Figure 2**. We will use this graph to illustrate the terminologies and notations that are introduced.

- The metric representation of node v_i is denoted by $\mathbf{d}_{v_i} = (dist_{v_1,v_i}, dist_{v_2,v_i}, \dots, dist_{v_n,v_i})$.
  - For example, in **Figure 2**, $\mathbf{d}_{v_1} = (0, 1, 2, 3, 3, 2)$
- The diameter of G is the length of the longest shortest path and is denoted by $diam(G) = \max_{v_i,v_j \in V}\left\{dist_{v_i,v_j}\right\}$.
  - For example, in **Figure 2**, $diam(G) = 3$.
- The open neighborhood of node v_i is a subset of all nodes directly connected to v_i and denoted by $\mathsf{Nbr}(v_i) = \{v_j | \{v_i, v_j\} \in E\}$.
  - For example, in **Figure 2**, $\mathsf{Nbr}(v_2) = \{v_1, v_3, v_6\}$.
- The metric representation of a node v_i with respect to a subset such as $S \subset V$ is denoted by $\mathbf{d}_{v_i,-S}$.
  - For example, in **Figure 2**, $\mathbf{d}_{v_1,-\{v_3,v_4\}} = (2, 3)$.

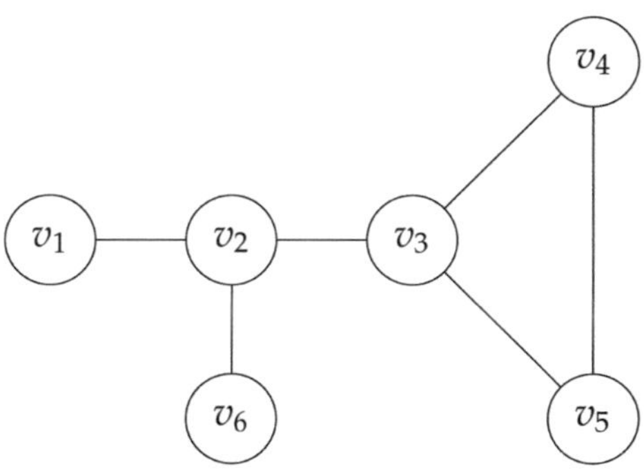

	v_1	v_2	v_3	v_4	v_5	v_6
v_1	0	1	2	3	3	2
v_2	1	0	1	2	2	1
v_3	2	1	0	1	1	2
v_4	3	2	1	0	1	3
v_5	3	2	1	1	0	3
v_6	2	1	2	3	3	0

Figure 2.
An example used in Section 3 for illustrating various notations.

- We can expand the previous notation to reflect the metric representation of a subset of nodes $V' \subset V\ S$ with respect to S as $\mathcal{D}_{V',-S} = \{\mathbf{d}_{v_l,-S} \mid v_l \in V'\}$.
  - For example, in **Figure 2**, $\mathcal{D}_{\{v_1,v_2\},-\{v_3,v_4\}} = \{(2,3),(1,2)\}$. Note that the first pair (2,3) corresponds to v_1 and the second pair (1,2) corresponds to v_2.
- We define a partition $\prod = \{V_1,\ V_2,\ \dots,V_t\}$ of $V' \subseteq V$ as one with the following properties:
  - $\bigcup_{i=1}^{t}\ V_i = V'$, and
  - for all $i \neq j$, $V_i \cap V_j = \emptyset$.
- We define a refinement $\prod' = \{V'_1, V'_2, \dots, V'_\ell\}$ of a partition $\prod$, denoted by $\prod' \prec_r \prod$, as one that can be obtained from $\prod$ using the following rules:
  - For every node $v_j \in \left(\bigcup_{i=1}^{t}\ V_i\right) \setminus \left(\bigcup_{i=1}^{\ell}\ V'_i\right)$, remove v_j from the set in $\prod$ that contains it.
  - Optionally, for every set V_ℓ in $\prod$, replace V_ℓ by a partition of V_ℓ.
  - If there exists an empty set, remove it.

    i. For example, in **Figure 2**, $\{\{v_1,v_2\},\{v_3\},\{v_5\}\} \prec_r \{\{v_1,v_2\},\{v_3,v_4,v_5\}\}$.
- We define an *equivalence relation* (and related notations) over set of same-length vectors $\mathcal{D}_{V\setminus V',-V'}$ for some $\emptyset \subset V' \subset V$ as follows:
  - The set of equivalence classes, which forms a partition of $\mathcal{D}_{V\setminus V',-V'}$, is denoted by $\prod^{\equiv}_{V\setminus V',-V'}$

    i. For example, in **Figure 2**, $\prod^{\equiv}_{\{v_1,v_2,v_6\},-\{v_3,v_5\}} = \{(2,3),(1,2),(2,3)\}$.
  - We declare two nodes $v_i, v_j \in V\setminus V'$ to be in the same equivalence class if $\mathbf{d}_{v_i,-V'}$ and $\mathbf{d}_{v_j,-V'}$ belong to the same equivalence class in $\prod^{\equiv}_{V\setminus V',-V'}$; thus $\prod^{\equiv}_{V\setminus V',-V'}$ also defines a partition into equivalence classes of $V\setminus V'$.
  - The *measure* of the equivalence relation is defined as

    $\mu\left(\mathcal{D}_{V\setminus V',-V'}\right) \stackrel{\text{def}}{=} \min_{y \in \prod^{\equiv}_{V\setminus V',-V'}} \{|y|\}$.
  - If a set S is a k-antiresolving set then $\mathcal{D}_{V\setminus S,-S}$ defines a partition into equivalence classes of measure k.

    i. For example, in **Figure 2**, $\mu\left(\mathcal{D}_{\{v_1,v_2,v_6\},-\{v_3,v_5\}}\right) = 1$ and $\{v_3,v_5\}$ is a 1-antiresolving set.

4. Theoretical results

To understand graph resistance against privacy attacks, one needs to study the (k, ℓ)-anonymity in greater details. Thus, we look into some computational problems related to this measure that were formalized and investigated in [16]. This section contains three problems from [16] and the respective algorithms to solve each problem efficiently. It is important to note that (k, ℓ)-anonymity in its basic definition sets no limitation for the adversary, which means that an adversary can take control of as many nodes as she/he can. However, in real world there are many mechanisms designed solely to prevent such attacks and thus the chances of being caught are significantly high. This notion is the motivation behind several problems with respect to measuring the (k, ℓ)-anonymity in a graph [17].

We now state the three problems for analyzing (k, ℓ)-anonymity. Problem 1 simply checks to find a k-antiresolving set for the largest possible value of k. Problem 2 sets a restriction for number of nodes the adversary can control and attempts to find the largest possible value of k while minimizing the number of nodes that are compromised. Problem 3 introduces a version of the problem that attempts to address the trade-off between privacy threshold and number of compromised nodes.

Problem 1 (metric antidimension ($ADIM$)). Find a k-antiresolving subset of nodes S that maximizes k.

Problem 1 assumes there are *no* limitations on the number of attacker nodes, thus finding an absolute bound for privacy violation. Note that solution to Problem 1, denoted by k_{opt}, shows that, given no bound on number of the nodes an adversary can control, it is feasible to uniquely re-identify k_{opt} nodes with probability $\frac{1}{k_{opt}}$. The assumptions in Problem 1 are rarely plausible in practice; due to mechanisms present to counter such attacks, the more nodes the adversary controls, the higher the risk of being exposed. Thus, a limit on number of attacker nodes is necessary, which leads us to Problem 2.

Problem 2 ($k_{\geq}$-metric antidimension ($ADIM_{\geq k}$)). Given k, find a k'-antiresolving set S such that (i) $k' >= k$ and, (ii) S is of minimum cardinality.

Problem 2 is an extension to Problem 1 that attempts to find the largest value of k while minimizing the number of attacker nodes. A solution to this problem asserts few interesting statements. For example, an adversary controlling l attacker nodes where $\ell < |\mathcal{L}_{opt}^{\geq k}|$ cannot uniquely re-identify any node in the network with a probability better than $\frac{1}{k}$. However, using enough number of nodes ($\geq |\mathcal{L}_{opt}^{\geq k}|$) one can re-establish such possibilities.

The third problem focuses on a trade-off between number of attacker nodes and the privacy violation probability. Given two measures (k, ℓ)-anonymity and (k', ℓ')-anonymity where $k' > k$ and $\ell' < \ell$, it is easy to observe that (k', ℓ')-anonymity measure provides a smaller privacy violation probability but also has lower tolerance for attacker nodes. The trade-off leads us to the third problem.

Problem 3 (k=-metric antidimension ($ADIM_{=k}$)) Given a positive integer k, find a k antiresolving subset of nodes S with minimum cardinality if such a subset exists.

Chatterjee et al. [16] investigated Problems 1–3 from a computational complexity perspective. The following theorems summarizes their finding on Problems 1–3. The non-trivial mathematical proofs for these theorems are unfortunately outside of the scope of this chapter; we strongly recommend readers who are interested in the proofs to read the original paper [16].

Theorem 1. [16]

1. Both *ADIM* and $ADIM_{\geq k}$ can be solved in $O(n^4)$ time.

2. Both *ADIM* and $ADIM_{\geq k}$ can also be solved in $O\left(\frac{n^4 \log n}{k}\right)$ time with high probability.

Theorem 2. [16]

1. $ADIM_{=k}$ is NP-Complete for any k in the range $1 \leq k \leq n^{\varepsilon}$ where $0 \leq \varepsilon < \frac{1}{2}$ is any arbitrary constant, even if the diameter of the input graph is 2.

2. Assuming NP $\not\subseteq$ DTIME$\left(n^{\log \log n}\right)$, there exists a universal constant $\delta > 0$ such that $ADIM_{=k}$ does not admit $\left(\frac{1}{\delta} \ln n\right)$ approximation for any integer k in the range $1 \leq k \leq n^{\varepsilon}$ for any constant $0 \leq \varepsilon < \frac{1}{2}$, even if the diameter of the input graph is 2.

3. If $k = n - c$ for some constant c then $ADIM_{=k}$ can be solved in polynomial time.

Theorem 3. [16]

1. $ADIM_{=1}$ admits $(1 + \ln(n - 1))$ approximation in $O(n^3)$ time.

2. If G has at least one node of degree 1 then $ADIM_{=1}$ can be solved in $O(n^3)$ time.

3. If G does not contain a cycle of 4 edges then $ADIM_{=1}$ can be solved in $O(n^3)$ time.

4.1 Algorithms

The following algorithms were devised in [16] to address Problems 1–3. It is important to note that *ADIM* can be solved in $O(n^5)$ time by repeatedly solving $ADIM_{\geq k}$ for $k = n - 1, n - 2, \ldots, 1$ to find the largest obtainable value for k such that $\mathcal{L}_{opt}^{\geq k} < \infty$. However, few modifications to Algorithm 1 directly result in $O(n^4)$ solution, which is shown in Algorithm 2.

5. Empirical results

In [18], DasGupta et al. investigated the resistance of 8 real-world network against active attacks with respect to the (k, ℓ)-anonymity. All the networks under investigation were unweighted graphs and the direction of edges (if the network was directed) was ignored during the analysis. **Table 1** contains the general information regarding these networks. Results for both *ADIM* and $ADIM_{\geq k}$ were obtained by running Algorithm 1 on the networks, the return statements from Algorithm 1 being an exact solution to Problem 2. On the other hand, the exact solution for Problem 1 can be achieved by combining Algorithm 1 and binary search on k to find the largest value of k such that $V_{opt}^{\geq k} \neq \emptyset$ [18].

Algorithm 1: $O(n^4)$ time deterministic algorithm for $ADIM_{\geq k}$ [16]

1 Compute $\mathbf{d}_i$ for all $i = 1, 2, \ldots, n$ in $O(n^3)$ time using Floyd-Warshall algorithm [17]
2 $\widehat{\mathcal{L}_{opt}^{\geq k}} \leftarrow \infty$ $\widehat{V_{opt}^{\geq k}} \leftarrow \emptyset$
3 **for** *each* $v_i \in V$ **do**
4     $V' = \{v_i\}$; done $\leftarrow$ FALSE
5   **while** $(V \setminus V' \neq \emptyset) \wedge (\neg\, done)$ **do**
6   Compute $\mu(\mathcal{D}_{V \setminus V', -V'})$
7   **if** $\mu(\mathcal{D}_{V \setminus V', -V'} \geq k$ **and** $|V'| < \widehat{\mathcal{L}_{opt}^{\geq k}}$ **then**
8             $\widehat{\mathcal{L}_{opt}^{\geq k}} \leftarrow V'$; $\widehat{V_{opt}^{\geq k}} \leftarrow V'$; done $\leftarrow$ TRUE
9   **else**
10   let $V_1, \ldots, V_l$ be the only $\ell > 0$ equivalence classes
11   in $\prod =_{V \setminus V', -V'}$ such that
12             $|V_1| = \ldots = |V_l| = \mu(\mathcal{D}_{V \setminus V', -V'})$
13             $V' \leftarrow V' \cup (\cup_{t=1}^{\ell} V_t)$
14   **end**
15   **end**
16 **end**
17 **return** $\widehat{\mathcal{L}_{opt}^{\geq k}}$ *and* $\widehat{V_{opt}^{\geq k}}$ *as our solution*

Algorithm 2: $O(n^4)$ time deterministic algorithm for *ADIM* [16]

1 Compute $\mathbf{d}_i$ for all $i = 1, 2, \ldots, n$ in $O(n^3)$ time using Floyd-Warshall algorithm [17]
2 $\widehat{V_{opt}^{\geq k}} \leftarrow \emptyset$; $\widehat{k_{opt}} \leftarrow 0$
3 **for** *each* $v_i \in V$ **do**
4     $V' = \{v_i\}$
5   **while** $V \setminus V' \neq \emptyset$ **do**
6   compute $\mu(\mathcal{D}_{V \setminus V', -V'})$
7   **if** $\mu(\mathcal{D}_{V \setminus V', -V'}) > \widehat{k_{opt}}$ **then**
8             $\widehat{k_{opt}} \leftarrow \mu(\mathcal{D}_{V \setminus V', -V'})$
9             $\widehat{V_{opt}^{\geq k}} \leftarrow V'$
10   **else**
11   let $V_1, \ldots, V_\ell$ be the only $\ell > 0$ equivalence classes
12   in $\prod^{=}_{V \setminus V', -V'}$ such that
13             $|V_1| = \ldots = |V_l| = \mu(\mathcal{D}_{V \setminus V', -V'})$
14             $V' \leftarrow V' \cup (\cup_{t=1}^{\ell} V_t)$
15   **end**
16   **end**
17 **end**
18 **return** $\widehat{k_{opt}}$ *and* $\widehat{V_{opt}^{\geq k}}$ *as our solution*

Algorithm 3: (*resp.* **Algorithm 4**) $O(\frac{n^4 logn}{k})$ time randomized algorithm for $ADIM_{\geq k}$ (*resp.* ADIM) [16]

1 Compute $\mathbf{d}_i$ for all $i = 1, 2, \ldots, n$ in $O(n^3)$ time using Floyd-Warshall algorithm [17]
2 $\widehat{\mathcal{L}_{opt}^{\geq k}} \leftarrow \infty$ $\widehat{V_{opt}^{\geq k}} \leftarrow \emptyset$ (for $ADIM_{\geq k}$)
    or
    $\widehat{V_{opt}^{\geq k}} \leftarrow \emptyset$; $\widehat{k_{opt}} \leftarrow 0$ (for *ADIM*)
3 **repeat**
4   Select a node v_i uniformly at random from the n nodes
5   execute step 4 to step 15 of Algorithm 1 (for $ADIM_{\geq k}$)
        or
    execute step 4 to step 16 of Algorithm 2 (for *ADIM*)
6 **until** $\lceil \frac{2n \ln n}{k} \rceil$ *times*
7 **return** *the best of all solutions found in the previous steps*

Algorithm 4: $O(n^3)$ time $(1 + ln(n - 1))$-approximation algorithm for $ADIM_{=1}$ [16]

```
Compute d_i for all i = 1, 2, ..., n in O(n^3) time using Floyd-Warshall
  algorithm [17]
L̂^{=1}_{opt} ← ∞ ; V̂^{=1}_{opt} ← ∅
(** Guessing that set {v_i} belongs to Π^{=}_{V\V^{=1}_{opt}, −V^{=1}_{opt}} **)
for each v_i ∈ V do
  | create an instance of standard set cover problem containing n − 1
  |   elements and n − 1 sets:
  | 𝒰 = { a_{v_j} | v_j ∈ V \ {v_j}}
  | S_{v_j} = { a_{v_j} } ∪ { a_{v_l} | dist_{v_i,v_j} ≠ dist_{v_l,v_j} } for j ∈ {1, 2, ..., n} \ {i}
  | if ∪_{j∈{1,2,...,n}\{i}} S_{v_j} = 𝒰 then
  |   | run the greedy approximation algorithm in [19] for the instance of
  |   |   the set cover problem giving a solution 𝒥 ⊆ {1, 2, ..., n} \ {i}
  |   | V′ = { v_j | j ∈ 𝒥}
  |   | if |V′| < ℒ̂^{=1}_{opt} then
  |   |   | ℒ̂^{=1}_{opt} ← |V′|
  |   |   | V̂^{=1}_{opt} ← V′
  |   | end
  | end
end
return ℒ̂^{=1}_{opt} and V̂^{=1}_{opt} as our solution
```

The results for both Problem 1 and Problem 2 for the networks in **Table 1** are depicted in **Table 2**. Results in **Table 2** provide the following interesting insights with respect to resistance against privacy attacks in real-world social networks [19].

- All networks, with the exception of "Enron Email Data" network, will have a significant percentage of their users compromised if an adversary gains control of *only* one node (varying between 2.6% of users compromised in "University Rovira i Virgili emails" network to 26.5% of users compromised in "Zachary Karate Club" network).

Name	Number of nodes	Number of edges	Description
Zachary Karate Club [20]	34	78	Network of friendship between 34 members of a karate club
San Juan Community [21]	75	144	Network for visiting relations between families living in farms in San Juan Sur, Costa Rica, 1948
Jazz Musician Network [22]	198	2842	A social network of jazz musicians
University Rovira i Virgili emails [23]	1133	10903	The network of email interchanges between members of university
Enron Email Data Set [24]	1088	1767	Enron email network
Email Eu Core [25]	986	24989	Emails from a large European research institute
UC Irvine College Message platform [26]	1896	59835	Messages on a Facebook-like platform at UC-Irvine
Hamsterster friendships [27]	1788	12476	Friendships between users of the website

Table 1.
List of 8 social networks studied in [18].

Name	n	k_{opt}	$p_{\text{opt}} = \frac{1}{k_{\text{opt}}}$	$\mathcal{L}_{\text{opt}}^{\geq k_{\text{opt}}} = \mathcal{L}_{\text{opt}}^{=k_{\text{opt}}}$	$\frac{k_{\text{opt}}}{n}$
Zachary Karate Club [20]	34	9	0.111	1	26.5%
San Juan Community [21]	75	7	0.143	1	9.3%
Jazz Musician Network [22]	198	12	0.084	1	6.0%
University Rovira i Virgili emails [23]	1133	29	0.035	1	2.6%
Enron Email Data Set [24]	1088	153	0.007	935	14.1%
Email Eu Core [25]	986	39	0.026	1	3.4%
UC Irvine College Message platform [26]	1896	55	0.019	1	2.9%
Hamsterster friendships [27]	1788	4	0.25	1	0.22%

n depict the number of nodes, k_{opt} is the largest value of k such that $V_{opt}^{\geq k} \neq \emptyset$, and $\mathcal{L}_{opt}^{\geq k_{opt}}$ is minimum number of attacker nodes for corresponding k.
[a]n denotes the number of nodes in the social graph.
[b]k_{opt} is the largest value of k such that $V_{opt}^{\geq k} \neq \emptyset$.

Table 2.
Results for ADIM *using Algorithm 1 [18].*

	k	4	5	10	20	40	60	100	120	153
Enron Email Data Set	$p_k = \frac{1}{k}$	0.25	0.2	0.1	0.05	0.025	0.017	0.01	0.009	0.007
	$\mathcal{L}_{opt}^{\geq k}$	1	334	463	567	683	842	935	935	935

Table 3.
$\mathcal{L}_{opt}^{\geq k}$ values recorded for $k > 1$ for the "Enron Email Data" network [18]. The values shown are subject to $\mathcal{L}_{opt}^{\geq k} \neq \mathcal{L}_{opt}^{\geq k-1}$.

- For all networks with the exception of "Enron Email Data" network, the minimum privacy violation probability is notably higher than 0 (varying between 0.019 for the "UC Irvine College Message platform" network to 0.25 for the "Hamsterster friendships" network). The value for minimum privacy violation probability in "Hamsterster friendships" network is notably higher compare to all other networks.

- In comparison to other networks, the "Zachary Karate Club" and the "San Juan Community" have higher percentage of their users compromised if subjected to a privacy attack.

The exception network is the "Enron Email Data" network which due to a high value of $\mathcal{L}_{opt}^{\geq k}$ is very resilient against an attack. An adversary needs to control at least 86% of the network to achieve a value of $p_{opt} = 0.007$, which is not feasible in practice. This interesting observation in the "Enron Email Data" network motivated further inspections in different values of k. As shown in **Table 3**, $\mathcal{L}_{opt}^{\geq k}$ in the "Enron Email Data" network does not decrease significantly until k is set to a much smaller value compare to k_{opt}, which further emphasizes that **violating the privacy of the "Enron Email Data" network is not guaranteed in practice.** The authors in [18] also investigated the (k, ℓ)-anonymity measure in *synthetic* networks constructed based on both Erdös-Rényi random graphs and Barabási-Albert scale-free networks. We refer the reader to the original paper for more information.

6. Conclusions

Since their emergence about a decade ago, social networks have rapidly grown and infiltrated every aspect of our daily lives. With rapidly expanding reliance on their platforms, social networks like Facebook and Twitter are becoming a goldmine of personal information and user behavior data which makes the study of these networks of prime importance. The valuable information stored within these platforms makes them the target of malicious entities which try to compromise the privacy of the users which may further lead to unwanted disclosure of the sensitive attributes of the network.

In this chapter, we have reviewed a novel privacy measure that quantifies the resistance of a large social network against a privacy violating attack. We reviewed some efficient algorithms to compute this measure in social graph and revisited the privacy violation properties in 8 real-world networks. The current theoretical and empirical results for (k, ℓ)-anonymity pave the way for further investigation of this measure, as well as addressing its shortcomings and limitations.

Acknowledgements

We thank Ismael G. Yero for useful discussions. This research was partially supported by NSF grants IIS-1160995 and IIS-1814931.

Author details

Tanima Chatterjee†, Nasim Mobasheri† and Bhaskar DasGupta*†
Department of Computer Science, University of Illinois at Chicago, Chicago, IL, USA

*Address all correspondence to: bdasgup@uic.edu

†These authors contributed equally.

References

[1] Trujillo-Rasua R, Yero IG. k-metric antidimension: A privacy measure for social graphs. Information Sciences. 2016;**328**:403-417

[2] Bar-Yehuda R, Chor B, Kushilevitz E, Orlitsky A. Privacy, additional information and communication. IEEE Transactions on Information Theory. 1993;**39**(6):1930-1943

[3] Comi M, DasGupta B, Schapira M, Srinivasan V. On communication protocols that compute almost privately. Theoretical Computer Science. 2012; **457**:45-58

[4] Feigenbaum J, Jaggard AD, Schapira M. Approximate privacy: Foundations and quantification. In: Proceedings of the 11th ACM Conference on Electronic Commerce; ACM. 2010. pp. 167-178

[5] Kushelvitz E. Privacy and communication complexity. SIAM Journal on Discrete Mathematics. 1992; **5**(2):273-284

[6] Yao AC. Some complexity questions related to distributive computing (preliminary report). In: Proceedings of the 11th Annual ACM Symposium on Theory of Computing; ACM. 1979. pp. 209-213

[7] Newman ME. The structure and function of complex networks. SIAM Review. 2003;**45**(2):167-256

[8] Albert R, Barabási AL. Statistical mechanics of complex networks. Reviews of Modern Physics. 2002; **74**(1):47

[9] Zhou B, Pei J, Luk W. A brief survey on anonymization techniques for privacy preserving publishing of social network data. ACM SIGKDD Explorations Newsletter. 2008;**10**(2): 12-22

[10] Backstrom L, Dwork C, Kleinberg J. Wherefore art thou r3579x?: Anonymized social networks, hidden patterns, and structural steganography. In: Proceedings of the 16th International Conference on World Wide Web; ACM. 2007. pp. 181-190

[11] Netter M, Herbst S, Pernul G. Analyzing privacy in social networks—An interdisciplinary approach. In: 2011 IEEE 3rd International Conference on Privacy, Security, Risk and Trust and 2011 IEEE 3rd International Conference on Social Computing; IEEE. 2011. pp. 1327-1334

[12] Wu X, Ying X, Liu K, Chen L. A survey of privacy-preservation of graphs and social networks. In: Managing and Mining Graph Data; Springer, Boston, MA. 2010. pp. 421-453

[13] Slater PJ. Leaves of trees. Congressus Numerantium. 1975;**14** (549-559):37

[14] Harary F, Melter RA. On the metric dimension of a graph. Ars Combinatoria. 1976;**2**(191-195):1

[15] Yu H, Gibbons PB, Kaminsky M, Xiao F. Sybillimit: A near-optimal social network defense against sybil attacks. In: 2008 IEEE Symposium on Security and Privacy; IEEE. 2008. pp. 3-17

[16] Chatterjee T, DasGupta B, Mobasheri N, Srinivasan V, Yero IG. On the computational complexities of three problems related to a privacy measure for large networks under active attack. Theoretical Computer Science. 2019; **775**:53-67

[17] Leiserson CE, Rivest RL, Cormen TH, Stein C. Introduction to Algorithms. Cambridge, MA: MIT Press; 2001

[18] DasGupta B, Mobasheri N, Yero IG. On analyzing and evaluating privacy

measures for social networks under active attack. Information Sciences. 2019;**473**:87-100

[19] Johnson DS. Approximation algorithms for combinatorial problems. Journal of Computer and System Sciences. 1974;**9**(3):256-278

[20] Zachary WW. An information flow model for conflict and fission in small groups. Journal of Anthropological Research. 1977;**33**(4):452-473

[21] Loomis CP, Morales JO, Clifford RA, Leonard OE. Turrialba: Social Systems and the Introduction of Change. Glencoe, IL: Free Press; 1953

[22] Gleiser PM, Danon L. Community structure in jazz. Advances in Complex Systems. 2003;**6**(04):565-573

[23] Guimera R, Danon L, Diaz-Guilera A, Giralt F, Arenas A. Self-similar community structure in a network of human interactions. Physical Review E. 2003;**68**(6):065103

[24] Enron email network. Available from: UC Berkeley Enron Email Analysis website http://bailando.sims.berkeley.edu/enron_email.html

[25] Paranjape A, Benson AR, Leskovec J. Motifs in temporal networks. In: Proceedings of the 10th ACM International Conference on Web Search and Data Mining; ACM. 2017. pp. 601-610

[26] Panzarasa P, Opsahl T, Carley KM. Patterns and dynamics of users' behavior and interaction: Network analysis of an online community. Journal of the American Society for Information Science and Technology. 2009;**60**(5):911-932

[27] Hamsterster friendships network dataset–KONECT, April 2017. Available from: http://konect.uni-koblenz.de/networks/petster-friendships-hamster)

Section 3

Applications

Chapter 9

Beyond Differential Privacy: Synthetic Micro-Data Generation with Deep Generative Neural Networks

Ofer Mendelevitch and Michael D. Lesh

Abstract

Recent advances in generative modeling, based on large scale deep neural networks, provide a novel approach for sharing individual-level datasets (micro-data) without privacy concerns. Unlike differential privacy, which enforces a specific query mechanism on data to ensure privacy, generative models can accurately learn the statistical patterns of such micro-data and then be used to generate "synthetic data" that accurately reflects these statistical patterns, yet contain none of the original data itself, and thus can be safely shared for analysis and modeling without compromising privacy. The successful application of these techniques to various industries including healthcare, finance, and autonomous vehicles is promising and results in continued investment in research and development of generative models in both academia and industry.

Keywords: generative models, synthetic data, deep neural networks, micro-data

1. Introduction

Differential privacy, created more than a decade ago, continues to play an important role in protecting privacy of micro-data while enabling statistical analysis. Initially applied by statistics agencies such as the US census bureau, it is now well recognized that, although useful for some applications, differential privacy comes with significant limitation (e.g., [1]).

To understand some of the limitations of differential privacy, consider the following:

- Differential privacy is defined around the concept of a *mechanism*; as such, it is not intended to create "sharable datasets," but instead allows a user (analyst) to submit various types of queries (via the defined *query mechanism*), requesting some kind of aggregate statistics, like summary statistics of the original data. This limits the usability of differential privacy to queries that are supported by that mechanism.

- An appropriate *privacy budget* needs to be decided upon, and in practice it's often difficult to agree on what that budget needs to be. In fact, practical

IntechOpen

Figure 1.
Fake celebrity images created using generative modeling; none of these images are real people.

use-cases demonstrate that due to concerns about risk, most implementations end up with much higher budget than is necessary.

- Many mechanisms of differential privacy require noise to be added to the data in cases where the original data is highly skewed, resulting in reduced utility of the outputs, and in some cases rendering the whole exercise useless.

- In many specific fields of statistical analysis, users of micro-data are highly trained to use specific tools (STATA, SAS, R and Python) and query procedures, which often do not support the complexity of differential-privacy-protected mechanisms. This presents a behavior-change challenge whereby analysts need to be convinced to abandon their familiar methods and tools (which they may have been using for decades) in favor of the interactive system where the privacy-protected data is available.

Fortunately, deep generative models – a recent and novel approach in deep neural networks – provide an alternative for direct sharing of micro-data without privacy risk.

With generative models, a deep neural network algorithm uses the existing micro-data to approximate, with high accuracy, the underlying probability distribution of the data in some high-dimensional latent space. Once the probability distribution is approximated, the trained model can be used to generate any number of *synthetic* records by randomly sampling from that distribution. Those generated records are related to the original data only through the shared underlying probability distribution, and thus does not include any information that can be linked back to the original (private) records.

To further illustrate how synthetic data generation works, consider CelebA,[1] a dataset with more than 200,000 synthetic celebrity face images, each with 40 automatically extracted attribute annotations. Using generative models, researchers have demonstrated the ability to learn the underlying distribution well enough to generate photorealistic celebrity faces as is shown in **Figure 1** above.

[1] http://mmlab.ie.cuhk.edu.hk/projects/CelebA.html

This same technique can be applied to many other types of data – music, text, videos as well as healthcare, financial or insurance data. In this chapter we will explore synthetic data generation and its application, and how releasing synthetic micro-data can provide an alternative to differential privacy.

In Section 2, we explore synthetic data in more detail, and how generative models can create synthetic data. In Section 3, we discuss using variational auto-encoders as generative models, followed by Section 4, where we discuss generative adversarial networks. In Section 5, we discuss the application of generative models to healthcare data, and in Section 6, we discuss privacy in the context of synthetic data, and some approaches that combine differential privacy with synthetic data generation. Section 7 is a summary and discussion on future directions in synthetic data generation.

2. Generative models for synthetic data

Generative models are a class of mathematical models that approximate a probability distribution of some dataset and can be used to generate samples of data according to the modeled (or approximated) distribution. Such generated data is often called "synthetic data," "fake data," or "realistic but not real."

For a given data domain, consider a dataset A with N data records. For most practical cases, the dataset can be assumed to be drawn from some (usually unknown) probability distribution P(x). A *synthetic* dataset S is a dataset similar to A in terms of fields or structure, where records in S are randomly drawn from some probability distribution Q(x).

In an ideal world where Q(X) = P(X) we can clearly use S for various purposes of analysis and modeling, because they are sampled from the same distribution. The key idea behind generating synthetic data is as follows: can we accurately estimate this probability distribution P(x), such that $Q(X) \approx P(X)$, with high fidelity?

Let us look at a simple example – consider a one-dimensional series of values A, where A is drawn from a normal (Gaussian) distribution with mean μ and standard deviation σ. In other words, we know in this case that P(x) is the normal distribution with $P_{\mu,\sigma}(x) = \frac{1}{\sigma\sqrt{2\pi}}e^{-\frac{1}{2}\left(\frac{x-\mu}{\sigma}\right)^2}$, and that the values in A should fit this distribution. We can then use Gaussian fitting to estimate the values of μ and σ from the data, as is demonstrated in **Figure 2**.

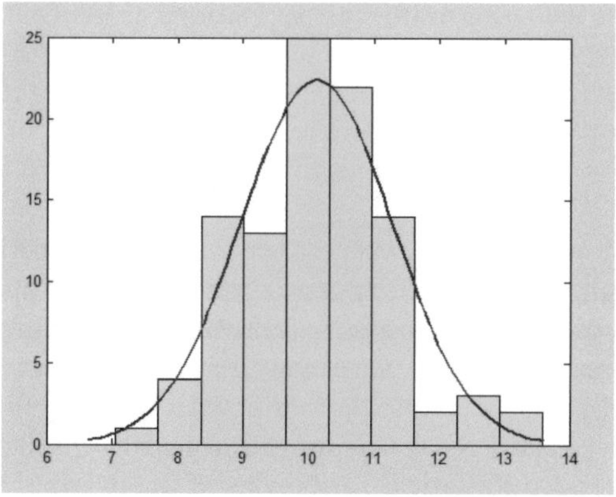

Figure 2.
Sample Gaussian fitting.

Once we have a good approximation for the parameters of the distribution (μ and σ), we can sample from this distribution to generate completely new data points that are fully consistent with the Gaussian distribution describing the original data.

This is of course a simplified example for two reasons. First, with a real generative model we do not know the actual form of the distribution function (e.g., Gaussian in this case); instead we use the neural network to estimate that function. Second, in the real world the data is not one-dimensional, but of much higher dimension.

So how do we approximate an unknown probability distribution from high-dimensional data?

The traditional approach to approximating data distribution is simple frequency counting (histograms), but of course this approach does not work in high dimensions due to the curse of dimensionality, namely the fact that most statistical methods fail in high-dimensional data due to increasing sparseness. This is also the case here with frequency counting, where with many dimensions the amount of histogram needed quickly explodes to make the method unfeasible.

Instead, the approach used in modern generative modeling research is to assume a functional form of the distribution $P_\theta(x)$ and learn the parameters θ of the function from the data. This set of parameters θ is in essence a compressed representation for the original dataset, often called "latent space representation."

To further illustrate this, let us go back to our example of celebrity images. Assume that the images are black and white (so that each pixel is represented by either 0 or 1), and of size $28 \times 28 = 784$ pixels. If we represent each image as a vector of 784 binary values, the number of possible values for a vector in this space is $2^{784} = 10^{236}$; if we want to approximate P(x) for each possible vector x in this space, we would need to estimate it for 10^{236} such vectors, which is clearly not realistic in practice (thus "the curse of dimensionality"). Instead, we can define some $P_\theta(x)$ with a much smaller set of parameters θ and estimate those parameters in such a way that $P_\theta(x) \approx P(x)$. It turns out that deep neural networks are a good match for this kind of problem, and can be used to accurately estimate the parameters of the distribution $P_\theta(x)$; there are many possible neural network architectures suitable for this task, most common of which are auto-encoders and generative adversarial networks.

Images are a very vivid (pun intended) demonstration of the power of generative models and how they can generate high utility synthetic data; but these techniques can also be successfully applied to many other fields such as music, poetry, cartoon characters, or even synthetic "video miles" for self-driving cars.

The performance of recent techniques in generative modeling is quite impressive, and their success led to a growth in applications of generative models in industry. For example, self-driving car companies use synthetic data to significantly increase the size of training data they have available, covering many more scenarios and edge-cases for improving their self-driving algorithms.

The usefulness of synthetic data generally falls into one of 3 important categories:

- **Replacement**. If access to the real dataset is limited or restricted (e.g., when data access is highly regulated), synthetic data often provides an excellent alternative. A good example comes from healthcare – access to medical records is often heavily restricted because of personal identifiers and the risk of linkage attacks. Synthetic medical records with high fidelity can provide the medical and bio-pharma research community with a replacement dataset that accurately reflects the statistical properties of the original data. This opens up an enormous opportunity to share and aggregate medical data from various clinical care sources and unlock important insights such as how effective are various therapeutics like drugs, medical devices or clinical care protocols.

- **Augmentation**. In many predictive modeling use-cases, the dataset available for training the model is relatively small in size, which often results in lower accuracy of the model. This phenomenon is further exacerbated when using deep learning for predictive modeling, where small datasets tend to overfit quite easily. Creating synthetic training examples and combining the real and synthetic data points ("augmenting" the real dataset with synthetic data), resulting in a much larger training dataset overall, can significantly improve the accuracy of the predictive models.

- **Equalization/reshaping**. An interesting aspect of using generative models is that we can generate as much data as is desired; often many more records than exist in the original dataset. A key characteristic of generative models is that we can direct them to shape the output dataset to certain desired criteria. For example, if the original dataset has 60% male and 40% female, we can control the gender distribution and generate a 50%/50% synthetic dataset. This enables users of the synthetic data to battle bias in the original dataset.

Equipped with a basic understanding of what synthetic data is, and how it's created using generative models, let us look in more detail at two of the most common types of generative models: variational auto-encoders and generative adversarial networks.

3. Variational auto-encoders

An autoencoder is a specific type of deep learning architecture which is split into two distinct neural networks: one is called the "encoder" and the other "decoder," as is shown in **Figure 3**.

In this architecture, the encoder E_θ is a deep neural network that encodes the input data (X) into some intermediate representation (Z, often referred to as "latent representation") in a reduced dimensional space, and the decoder D_θ is also a deep neural network that decodes the vector Z back into the output vector Y. X and Y are of the same dimensionality. The goal of training the auto-encoder is to reconstruct the input X in the output Y, while transitioning through the lower dimensionality representation Z, so that we get as close as possible to $Y = D_\theta(E_\theta(X))$. If you optimize this auto-encoder in such a way that the loss of data between input X and

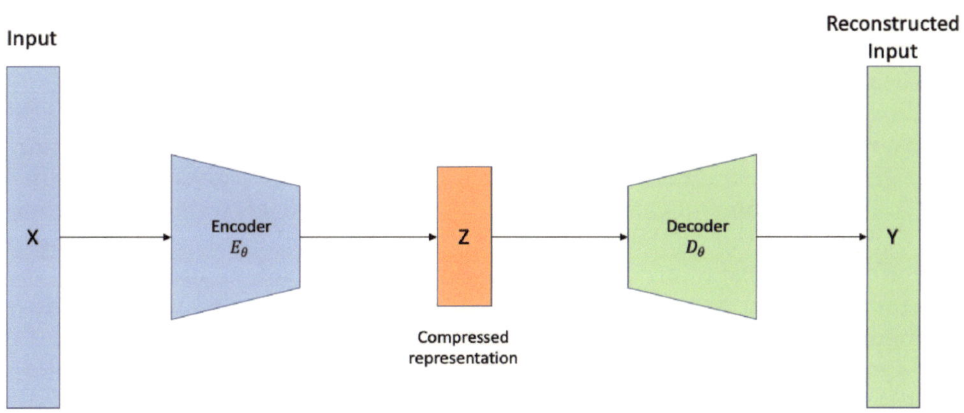

Figure 3.
Auto-encoder architecture.

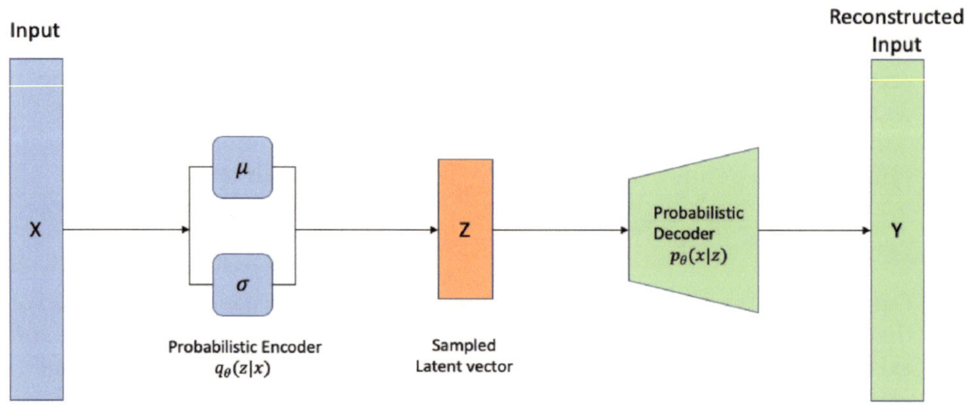

Figure 4.
Variational auto-encoder architecture.

output Y (reconstruction error) is minimized, then it's as if you are trying to find an optimal compressed representation for the input data.

Traditional auto-encoders have been around since the early days of neural networks and in their basic form they cannot be used to generate synthetic data; In 2013 the idea of *variational* auto-encoders (VAE) started to take shape, primarily with the work of [2, 3], as a way to use auto-encoders as generative models.

With VAEs, instead of mapping the input vector X to a fixed vector Z, we want to map it into a distribution $q_\theta(z|x)$, often assumed to be a multivariate normal distribution with mean μ and standard deviation σ; then to generate synthetic outputs Y we just randomly sample this learned distribution and decode the sampled vector to arrive at a synthetic output Y, as shown in **Figure 4**.

VAEs, being one of the first deep neural network architectures for practical generative models, created a lot of excitement about synthetic data, and was used primarily to generate synthetic images. Although elegant and theoretically pleasing, the synthetic images generated by VAEs tend to be blurry, which very quickly became a limiting factor for their use in synthetic imaging. Various improvements to the basic VAE approach have been proposed such as beta-VAE [4] and VQ-VAE [5] to address these issues; however, this also led researchers to the idea of generative adversarial networks, which we discuss next.

4. Generative adversarial networks

The idea of a generative adversarial network is inspired by game theory: we build two models, a generator and a discriminator, that compete with each other in an adversarial manner to collaboratively optimize the whole system. The generator G is a generative neural network that outputs synthetic samples given a noise variable Z. The discriminator D is a different neural network that is trained to discriminate between real and synthetic samples. During training, the generator is trying to generate samples that mimic as much as possible the real data, so that it can fool the discriminator, whereas the discriminator is trained not to be fooled and be able to distinguish between real and synthetic data samples. This is shown in **Figure 5**.

As you can see from **Figure 5**, a key idea in this architecture is that the discriminator D shares gradient updates with the generator, such that the generator can "understand" how its generated data fails to fool the discriminator and improve its generation over time resulting in better and better synthetic samples.

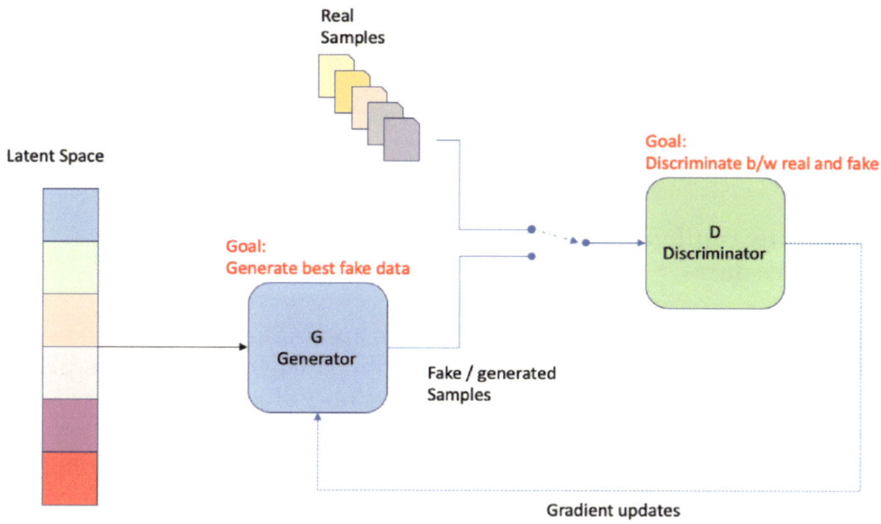

Figure 5.
Generative adversarial networks.

GANs were first formulated by Ian Goodfellow and colleagues [6], and since then have been an active area of research; they have demonstrated the ability to generate significantly better synthetic images than VAEs, and have been used in a variety of applications like generating synthetic celebrity faces, fake Pokémon characters, time-series medical events [7] and electronic medical records [8].

Due to the impressive realism in synthetic data generated by GANs, they have also initiated an active and important discussion of malicious use of generative models, and privacy implications. We will discuss this important aspect of generative models in Section 6.

One difficulty with GANs is that they are quite difficult to train, and often require significant time and effort to manually tune until they reach the desired outcome; some of the most common issues when training GANs are:

- **Nash Equilibrium:** the Generator and Discriminator work against each other in a competitive manner, and it is often rather difficult to reach the Nash equilibrium of this 2-player minimax game. Training GANs to achieve this equilibrium tends to require extensive experimentation and good intuition about how GANs work.

- **Vanishing gradient:** when the Discriminator is doing very well in its role to discriminate between real and synthetic data, its gradients are very close to 0 and thus learning in the Generator slows down significantly or sometimes even stops completely.

- **Mode collapse:** a common failure mode in GANs where the Generator generates samples that "fool" the Discriminator but fails to generate the full breadth of such possible samples and thus gets stuck in a local sub-space of synthetic samples possible. For example, consider an image face generator that generates excellent photorealistic images of faces but only focuses on faces of people with gray hair. Since the images are of great quality, the discriminator will consider them of great quality and indistinguishable from real images, however they only represent a fraction of the types of images in the training set, which include many more hair colors.

Various approaches and hacks have been proposed to address the vulnerabilities in GANs with varying levels of success. One important improvement over the basic GAN approach is Wasserstein GAN (WGAN [9]) which uses a different loss function based on Wasserstein distance, and has been shown to be more robust to mode collapse.

5. Industry example: applications of generative models to healthcare data

Healthcare is one of the most popular area of application for analytics and machine learning, driving improved outcomes for patients, lower cost of care, and improved patient experience. There are a vast number of applications for data in healthcare, such as measuring quality of care metrics, developing predictive models for better diagnosis, or analyzing data to understand the differences in clinical care protocols.

Due to the highly regulated nature of healthcare data, and various regulations that govern health data privacy (such as HIPAA, GDPR, CCPA), most healthcare data are locked down in silos. Many healthcare organizations have used de-identification as a way to reduce privacy risks, typically through the modification of potentially identifiable attributes (e.g., dates of birth) via generalization, suppression or randomization. However, this approach is susceptible to linkage attacks, as was demonstrated in [10], and it is accepted by many risk experts that the risk of re-identification is high and in fact they treat de-identified medical data the same way they do fully identifiable medical data.

This presents an enormous challenge to realizing the promise of understanding and using data in healthcare to drive better outcomes and achieving the vision of precision medicine.

There are many types of medical data that is useful, and herein we focus on three types of data that are quite common:

- **Tabular data:** large amounts of medical data are collected in table format, including clinical trial data and other data used for observational studies. In clinical trials, for example, the researchers review the individual patient records from the trial, and perform statistical analysis to understand whether the hypothesized outcome of the trial is confirmed or rejected with statistical significance given the data. Being able to share the vast amount of clinical trial data that is currently locked down in medical centers and biopharma companies to the research community, as well as combining these datasets, can unlock advances in design and speed-to-market for many necessary drugs and medical devices.

- **Electronic Medical Records:** electronic medical records (EMR) are now mandated by regulatory bodies; a vast number of such records is collected every day around the world, and stored in EMR systems by vendors like EPIC, Cerner and Allscripts. EMR are difficult to access due to privacy regulations, yet they represent a gold-mine of aggregated knowledge about health outcomes and can open up enormous opportunities for precision medicine.

- **Medical imaging**: medical imaging diagnostics using MRI, CT and other types of scanning are critical in diagnosis and following the response of treatment, and where advanced AI and machine learning are poised to provide significant gains in the near future (see e.g., [11]). Yet many diagnostics providers are starving for highly quality and diverse labeled medical images to improve their diagnostics models, leaving a huge gap in advancing the state of the art.

By providing synthetic EMR, clinical trial or medical imaging data that accurately mimics the statistical properties of the real data, one can perform the same analysis or modeling on the synthetic data, achieving near- identical results, without the risk of exposing patient privacy. Even more exciting is the ability to augment small medical datasets with synthetic data, which is useful for example in the case of relatively rare medical conditions where the number of patients available is limited.

It's interesting to note that there is previous work on synthetic data generation in the healthcare domain, notably the work done on Synthea described in [12]. These early techniques, while recognizing the importance of high fidelity synthetic data, used domain-specific knowledge to drive simulated data, but have unfortunately failed to achieve the kind of fidelity that is required for any meaningful analytics (see [13]), and thus have proven to be of limited use in practice where patient-level analysis is required.

More recently, generative adversarial networks and variational auto-encoders have been applied to medical datasets, which have demonstrated the potential to provide much higher fidelity synthetic data and thus more useful in practice. We now quickly review two of these more recent techniques: generating medical records with discrete values (MedGAN), and work by Nvidia to generate synthetic medical imaging.

5.1 MedGAN: generating discrete medical variables with GANs

Electronic medical records include vast amounts of structured data about patients such as diagnoses, drugs, lab results, and procedures. Most of this data is encoded in commonly shared data dictionaries such as ICD9 or ICD10 for diagnosis codes, NDC for drug codes, and similar dictionaries for procedure codes and labs. Although some variables in this data are continuous (like lab results), most of it is represented as discrete variables with very large dictionary sizes.

MedGAN [8] was developed with the recognition of the potential that generative adversarial networks have to model electronic medical records, while trying to adapt the GAN approach to deal with discrete variables, which it's not typically very good at. MedGAN aims to learn the probability distribution of data that include high-dimensional, multi-label discrete variables, and specifically supporting both binary (e.g., variables that represent whether you have a certain diagnosis or not), and count variables (i.e., variables that represent how many times a patient took a medication over time, or total number of risk factors for some disease). This approach proposes combining an auto-encoder within a generative adversarial network architecture and demonstrates how to deal with situations of overfitting and mode collapse in this scenario.

It is noteworthy that in addition to MedGAN, several researchers proposed additional similar approaches to modeling medical records and other tabular data, for example EhrGAN proposed in [14] and TableGan proposed in [15].

5.2 Medical image synthesis with GANs

It is widely recognized in AI and machine learning that insufficient data volume as well as imbalanced or non-diverse data often leads to poor predictive performance and lack of model generalization. This often proves to be a critical issue in the development of medical imaging algorithms where abnormal findings are by definition rare, and high-quality training images are hard to find.

In [16], Nvidia researchers demonstrate generation of synthetic MRI images with brain tumors using generative adversarial networks, trained on two publicly available datasets of brain MRI: ADNI and BRATS. Two distinct benefits of synthetic data are highlighted in this work: improved performance leveraging synthetic images as a form of data augmentation, and the value of synthetic data as a tool for reducing privacy risk while achieving comparable tumor segmentation results when trained on the synthetic data versus when trained on real data.

The results from [16] are quite impressive, and some synthetic images taken from that paper are shown in **Figure 6**.

Clearly more work remains in this area, especially in generating higher resolution synthetic images, tackling all imaging modalities as well as addressing many other clinical use-cases; nonetheless, this work demonstrates excellent initial results for synthetic image generation in medical research with the potential to improve medical imaging diagnostics and significantly reduce privacy risks.

5.3 Other approaches

Recently, neural language models with attention (i.e., Transformers [17]) have been used to for a variety of language tasks, including synthetic text generation, sequence to sequence translation, question answering and many others. One potential application of language models in medicine is the generation of free-text clinical notes based on structured data. Instead of generating synthetic versions of the structured medical EMR record, the goal is to translate the input structured data into a clinically correct and useful text summary of the patient information, in a form physicians are used to reading. Although early experiments with human-like language generation with models like GPT2 are showing good initial results, there's still a lot of work to do in this area.

It's worth mentioning one other generative modeling approach called flow-based generative models; this technique is quite complex mathematically, and is in early stages of research, but can potentially provide an additional set of

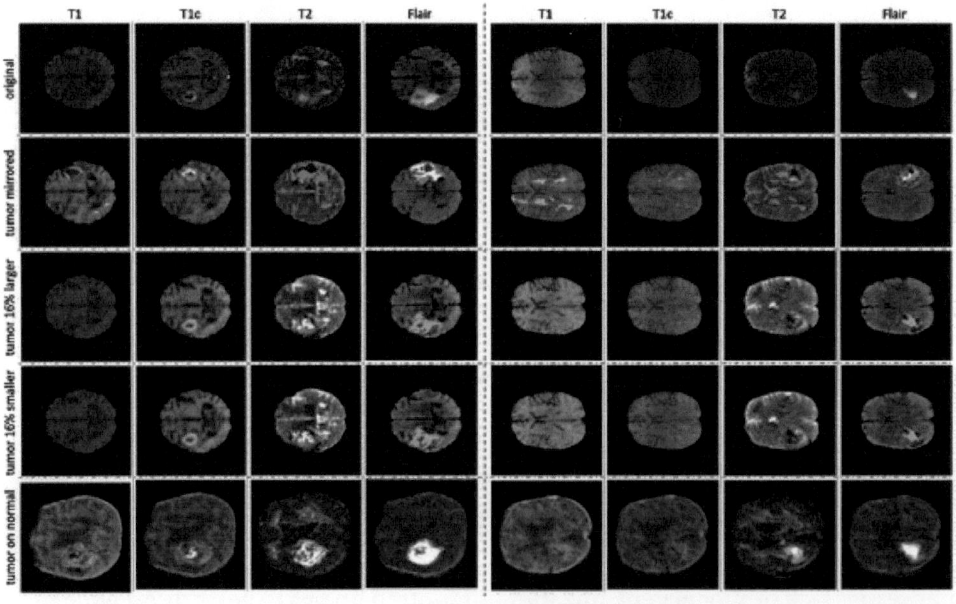

Figure 6.
Examples of synthetic abnormal brain MRI images.

methods for synthetic data generation. The interested reader is referred to [18, 19] for more details.

Another recent area of research in deep learning and privacy aims to integrate differential privacy into training procedures of deep neural networks [20]. This is particularly important for generative models and can be used to constrain the learning process around certain privacy guarantees, ensuring that the learning process does not just memorize the input data.

6. Privacy of synthetic data

With differential privacy, our goal is to define a query mechanism that guarantees certain privacy levels if the users are restricted to access micro-data through the specified mechanism only. Synthetic data generation is different in that it assumes synthetic data is published directly to users, and thus access to the data is virtually unlimited. We now want to inspect those differences in more detail to better understand the implications of privacy for synthetic data generation.

We start with an important, fundamental recognition. With real datasets (either de-identified or available through differential privacy mechanisms), an attacker knows for sure that each row in the datasets represents a real instance or person, only the privacy mechanisms attempt to conceal the privacy information in different ways. With synthetic datasets that is not the case, as the samples are randomly chosen from a probability distribution, and thus by definition do not reflect real people. In fact, as described at the beginning of this chapter, if we assume the real data and synthetic data are both sampled from a theoretical (unknown) distribution $P(X)$, and that distribution is very high dimensional (as it often is for micro-data), then the only hypothetical risk is that by a stroke of luck a synthetic record exactly matches the values in one of the original values, which is very unlikely. And its occurrence could not be recognized with any assurance by an attacker.

Nonetheless, there is an important privacy consideration – unintended memorization [21]. A deep generative learning model might unintentionally memorize the training set (of real data) and thus instead of approximating a distribution and then sampling from that distribution, it instead just copies one or more of the original data records into the synthetic dataset.

It is possible to test for memorization pro-actively as part of training the generative model (as proposed in [21]) and optimize the generative model in such a way as to remove any memorization or minimize it to a level which presents minimal risk.

To further enhance privacy guarantees, we can apply a k-anonymity [22] to the synthetic dataset. It's common to use generalization or obfuscation of variables to achieve the desired levels of k-anonymity; however both techniques result in reduced utility. With synthetic data, however, one can instead generate additional records in a way that improves the privacy guarantees without compromising utility.

7. Summary and conclusion

In this chapter we provided an overview of synthetic data and how it may provide an alternative to differential privacy as a method for sharing micro-data for the purpose of analysis and machine learning applications.

We discussed two of the most common techniques used in deep generative modeling, namely variational auto-encoders and generative adversarial networks, and highlighted some of the remarkable success in the space of modeling medical data. We then discussed why synthetic data provides privacy by design and some areas of research in privacy of synthetic data generation.

As research in the space of generative models continues at a neck-break pace at companies like OpenAI, Google, Facebook, Microsoft and others, we expect to see tremendous prosgress in this field on the research side as well as in applications of synthetic data across many areas of industry.

Acknowledgements

We would like to thank Dr. Johannes Otterbach and Dr. Marlene Grenon for many hours of discussions on the topic of synthetic data generation.

Conflict of interest

The authors are co-founders of Syntegra.io, a startup with a mission to enable completely secure data sharing of even the most sensitive medical information in a way that fully maintains statistical fidelity while preserving privacy, via its synthetic data engine based on generative models.

Author details

Ofer Mendelevitch[1*] and Michael D. Lesh[2]

1 Syntegra.io, San Carlos, USA

2 Syntegra.io, Mill Valley, USA

*Address all correspondence to: ofermend@gmail.com

References

[1] Garfinkel S, Abowd J, Powazek S. Issues encountered deploying differential privacy. In: WPES'18: Proceedings of the 2018 Workshop on Privacy in the Electronic Society; 2018. pp. 133-137. DOI: 10.1145/3267323.3268949

[2] Kingma D, Welling M. Auto-encoding variational bayes. In: ICLR; 2014. arXiv:1312.6114

[3] Rezende D, Mohamed S, Wierstra D. Stochastic backpropagation and approximate inference in deep generative models. In: Proceedings of the 31st International Conference on Machine Learning (ICML); 2014. arXiv:1401.4082v3

[4] Higgins I, Matthey L, Pal A, Burgess C, Glorot X, Botvinick M, et al. β-VAE: Learning basic visual concepts with a constrained variational framework. International Conference on Learning Representations. 2017;**2**(5):6

[5] Van den Oord A, Vinyals O, Kavukcuoglu K. Neural Discrete Representation Learning. In: NIPS; 2017. arXiv:1711.00937v2

[6] Goodfellow I, Pouget-Abadie J, Mirza M, Xu B, Warde-Farley D, Ozair S, et al. Generative adversarial nets. In: Advances in Neural Information Processing Systems. 2014. pp. 2672-2680. arXiv:1406.2661v1

[7] Yu L, Zhang W, Wang J, Yu Y. SeqGAN: Sequence generative adversarial nets with policy gradient. In: Proceedings of the Thirty-First AAAI Conference on Artificial Intelligence, February 4-9, 2017. San Francisco, California: AAAI Press; 2017. pp. 2852-2858. arXiv:1609.05473v6

[8] Choi E, Biswal S, Malin B, Duke J, Stewart W, Sun J. Generating multi-label discrete patient records using generative adversarial networks. In: Machine Learning for Healthcare Conference. PMLR; 2017. arXiv:1703.06490v3

[9] Arjovsky M, Chintala S, Bottou L. Wasserstein GAN; 2017. arXiv:1701.07875v3

[10] Barth-Jones D. The "Re-identification" of governor William Weld's medical information. In: A Critical Re-Examination of Health Data Identification Risks and Privacy Protections, Then and Now. 2012. DOI: 10.2139/ssrn.2076397

[11] Pesapane F, Codari M, Sardanelli F. Artificial intelligence in medical imaging: Threat or opportunity? In: Radiologists Again at the Forefront of Innovation in Medicine. 2018. DOI: 10.1186/s41747-018-0061-6

[12] Walonski J, Kramer M, Nichols J, Quina A, Moesel C, Hall D, et al. Synthea: An approach, method, and software mechanism for generating synthetic patients and the synthetic electronic health care record. Journal of the American Medical Informatics Association. 2018;**25**(3):230-238. DOI: 10.1093/jamia/ocx079

[13] Chen J, Chun D, Patel M, Chiang E, James J. The validity of synthetic clinical data: A validation study of a leading synthetic data generator (Synthea) using clinical quality measures. BMC Medical Informatics and Decision Making. 2019;**19**(1). DOI: 10.1186/s12911-019-0793-0

[14] Che Z, Cheng Y, Zhai S, Sun Z, Liu Y. Boosting deep learning risk prediction with generative adversarial networks for electronic health records. In: International Conference on Data Mining. IEEE; 2017. arXiv:1709.01648v1

[15] Park N, Mohammadi M, Gorde K, Jajodia S, Park H, Kim Y. Data synthesis

based on generative adversarial networks. In: International Conference on Very Large Data Bases. 2018. arXiv:1806.03384v5

[16] Shin H, Tenenholtz N, Rogers J, Schwartz C, Senjem M, Gunter J, et al. Medical image synthesis for data augmentation and anonymization using generative adversarial networks. In: Workshop on Simulation and Synthesis in Medical Imaging - SASHIMI2018. 2018. arXiv:1807.10225v2

[17] Vaswani A, Shazeer N, Parmar N, Uszkoreit J, Jones L, Gomez A, et al. Attention is all you need. In: NIPS17: Proceedings of the 31st International Conference on Neural Information Processing Systems. 2017. pp. 6000-6010. arXiv:1706.03762v5

[18] Dinh L, Sohl-Dickstein J, Bengio S. Density estimation using real NVP. In: ICLR. 2017. arXiv:1605.08803v3

[19] Kingma D, Dhariwal P. Glow: Generative flow with invertible 1x1 convolutions. In: Advances in Neural Information Processing Systems. 2018. pp. 10215-10224. arXiv:1807.03039v2

[20] Abadi M, Chu A, Goodfellow I, McMahan B, Mironov I, et al. Deep learning with differential privacy. In: Proceedings of the 2016 ACM SIGSAC Conference on Computer and Communications Security. ACM; 2016. pp. 308-318. arXiv:1607.00133v2

[21] Carlini N, Liu C, Erlingsson U, Kos J, Song D. The Secret Sharer: Evaluating and Testing Unintended Memorization in Neural Networks and Extracting Secrets; 2018. arXiv:1802.08232v3

[22] Holohan N, Antonatos S, Braghin S, Aonghusa P. (k, ε)-Anonymity: k-Anonymity with ε-Differential Privacy; 2017. arXiv:1710.01615v1

Chapter 10

Machine Learning Applications in Misuse and Anomaly Detection

Jaydip Sen and Sidra Mehtab

Abstract

Machine learning and data mining algorithms play important roles in designing intrusion detection systems. Based on their approaches toward the detection of attacks in a network, intrusion detection systems can be broadly categorized into two types. In the misuse detection systems, an attack in a system is detected whenever the sequence of activities in the network matches with a known attack signature. In the anomaly detection approach, on the other hand, anomalous states in a system are identified based on a significant difference in the state transitions of the system from its normal states. This chapter presents a comprehensive discussion on some of the existing schemes of intrusion detection based on misuse detection, anomaly detection and hybrid detection approaches. Some future directions of research in the design of algorithms for intrusion detection are also identified.

Keywords: misuse detection, signature detection, anomaly detection, hybrid detection, clustering, classification, unsupervised learning, supervised learning, intrusion detection system, network security

1. Introduction

Cyberinfrastructures are vulnerable to various possible attacks due to the flaws in their design and implementation. The major flaws that cause most of the critical vulnerabilities are errors in system programs and faulty design of the software. Malicious attackers can exploit these system vulnerabilities by following a sequence of activities, either from inside or from outside of the infrastructure, and cause significant damage. These events manifest themselves in the form of different distinct characteristics that are defined as patterns of attacks. Misuse or signature detection techniques attempt to proactively detect the presence of such patterns so that any malicious attack on the infrastructure can be effectively defended against. It is possible to defend against all known vulnerabilities in cyberinfrastructures by using supervised learning approaches for misuse and signature detection. The most convenient method of signature detection is measuring the similarity between the patterns recognized in the current network activity and the already known patterns of various types of cyber-attacks. However, execution signatures may vary substantially from one attack category to another, so that specific detection methods are required to classify attack patterns and, thus, to improve detection capability.

Anomaly detection systems, however, work in a different way. The objective of these systems is to proactively detect any activity or event in a network or host computer that exhibits aberration from the normal behavior of the network or the host. The normal behavior is described by a predefined set of activities. The working principle of an anomaly detection system is fundamentally different from that of misuse or signature detection system. Misuse or signature detection systems first need to be equipped with a well-defined set of attack signatures populated in their database. An anomaly detection system, on the other hand, defines a detailed and accurate profile of the normal behavior of the networks and hosts. The normal state of the cyberinfrastructure, consisting of networks and hosts, indicates an attack-free state. When an anomalous activity occurs in the cyberinfrastructure, the anomaly detection system notices a state change from the normal state to a state that is no longer normal. On observing this state change, the anomaly detection system raises an alert of a possible attack on the cyberinfrastructure. Unlike the signature or misuse detection systems, the anomaly detection systems are capable of detecting novel attacks as the detection strategy for these systems is based on the state change information, rather than a matching of attack signatures. It is precisely for this reason that anomaly detection schemes are capable of detecting various different types of attacks. Some of these attacks include: (i) segmentation of binary code in a user password, (ii) backdoor service on a malicious process on a well-known port number in a computing host, (iii) stealthy reconnaissance attempts, (iv) novel buffer overflow attacks, (v) direction of hypertext transmission protocol (HTTP) on a nonstandard port number, (vi) stealthy attacks on protocol stacks and (vii) different variants of denial of service (DoS) and distributed denial of service (DDoS), and so on. Early and accurate detection of these attacks poses significant challenges in the design of a robust and accurate anomaly detection system.

In this chapter, we have briefly reviewed some of the well-known misuse and anomaly-based detection systems that are proposed in the literature. We have also discussed some hybrid approaches in intrusion detections that effectively combine misuse and anomaly detection approaches so as to improve the detection accuracy and reduce false alarms. The rest of the chapter is organized as follows. Section 2 presents a brief discussion on misuse or signature-based detection approach. In Section 3, we discuss how various machine learning approaches can be applied in misuse or signature-based systems. Section 4 provides a brief overview of anomaly detection, while in Section 5, we discuss how machine learning and data mining algorithms can be effectively deployed in anomaly-based detection systems. In Section 6, we briefly discuss the working principles of some of the well-known hybrid detection systems. Section 7 concludes the chapter while highlighting some of the recent trends in machine learning approaches in network security applications.

2. Misuse or signature detection

Misuse detection, also called signature detection, is an approach in which attack patterns or unauthorized and suspicious behaviors are learned based on past activities and then the knowledge about the learned patterns is used to detect or predict subsequent similar such patterns in a network. The attack or misuse patterns, which are also called signatures, include patterns of log files or data packets that were found to be malicious and identified as threats to the network and the computing hosts. Each log file consists of its own signature that exhibits a unique pattern consisting of binary bits 0 and 1. For intrusion detection systems protecting host computers, that is, for host-based intrusion detection systems (HIDSs), the attack signature databases may contain various patterns of system calls that represent a different attack on the host. In the case of a network-based intrusion detection system (NIDS), attack signatures reveal specific patterns in data packets. These patterns may include signatures of the data payload, the packet header, unauthorized activities, such as improper file transfer protocol (FTP) initiation, or failed login attempt in Telnet. A typical data packet includes several fields such as: (i) the source Internet protocol (IP) address, (ii) the destination IP address, (iii) the source port number for transmission control protocol (TCP) or user datagram protocol (UDP), (iv) the destination port number for TCP or UDP, (v) the protocol description such as UDP, TCP or Internet control message protocol (ICMP), and (vi) the data payload. An attack signature can be detected in any specific field, or in any combination of these fields.

Figure 1 shows how a typical misuse or signature detection system works. These detection systems execute algorithms that attempt to match learned patterns or signatures from past attacks with the current activities in a network in order to detect any possible attack or malicious activities. If the signature of any current activity in the network matches with the signature of any activity in the attack signature database, the detection system raises an alert. A module in the detection system initiates a further investigation of the attack and starts invoking appropriate security modules to defend against such attacks. If the attack is found to be a real attack and not a false alarm by the detection system, the existing database of the attack signatures is updated with the signature of the new attack. For example, if the signature of an attack is: *login name* = "Sidra," then, whenever there is any attempt to login into any device in the network with the name "Sidra," the signature detection system will raise an alert of an attack.

This approach adopted in a signature-based detection system is primarily meant for detecting already known threats and vulnerabilities in a network. However, these systems suffer from a drawback of producing too many false alarms. A false alarm or a false positive refers to a situation where the system raises an alert of an attack while no attack has really happened on the network. As an example, let us consider the case where a user logs into a remote server. If the user forgets the login password and makes multiple attempts of login, the account of the user is most likely to be locked after a certain number of such failed attempts. As the signature-based detection system cannot differentiate between a failed login attempt by a legitimate user, and a malicious user attempting to login in an unauthorized way into some legitimate user's account, both the activities are considered as attacks.

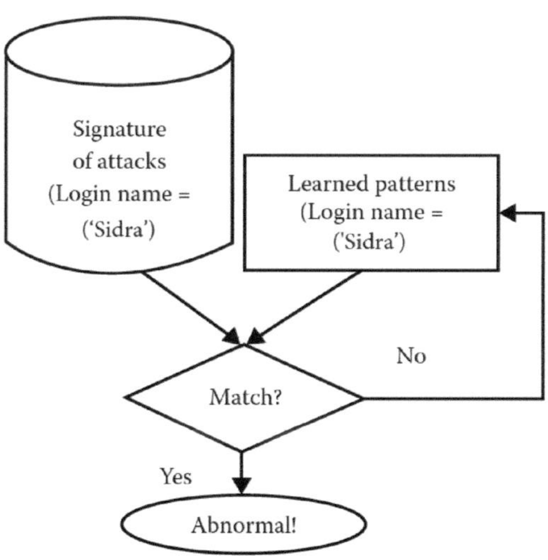

Figure 1.
Working of misuse or signature detection: Illustration of "if-else" rules.

The efficacy of misuse or signature detection system largely depends on the completeness and sufficiency of the knowledge of attack patterns and signatures captured in the attack signature database of the system. It is a nontrivial task to capture and represent the knowledge of attacks and system vulnerabilities in a cyberinfrastructure or in a network of computing machines, and the job heavily depends on domain experts. Since the knowledge and skills of domain experts may vary significantly from person to person, the design of signature detection systems, quite often, can be incomplete and inaccurate. Moreover, a slight variation, evolution, blending, or a combination of already known attacks can make signature detection an impossible task. This is a typical problem with any similarity-based learning system like a signature-based intrusion detection system.

3. Machine learning in misuse or signature detection

Figure 2 depicts the working mechanism of misuse or signature detection consists of five major steps: (i) data collection, (ii) data preprocessing, (iii) misuse or signature identification using a matching algorithm, (iv) rules regeneration and (v) denial of service (DoS) or other security response strategy. In most of the cases, the data sources are: network and host audit logs, packets transmitting over the network, and windows registry. Data preprocessing is a critical step that prepares the raw data for learning patterns. These steps involve the reduction of noise by eliminating outliers, normalizing or standardizing of data, and finally selecting and extracting features. After the data preprocessing step is over, an automatic intelligent learning system is deployed to build a learning model and extract rules using prior knowledge of the execution of malicious programs, network traffic data, and vulnerabilities in network infrastructure. The model is now ready for signature and misuse detection. The learned classification model is applied to the incoming network traffic for signature detection. If any part of the network traffic is found to be similar to attack patterns learned by the model, then an alarm is raised and the traffic is further analyzed for identifying whether it is really an attack or a false alarm. Consequently, misuse or signature detection can be simply understood as an "if-then" sequence as shown in **Figure 1**.

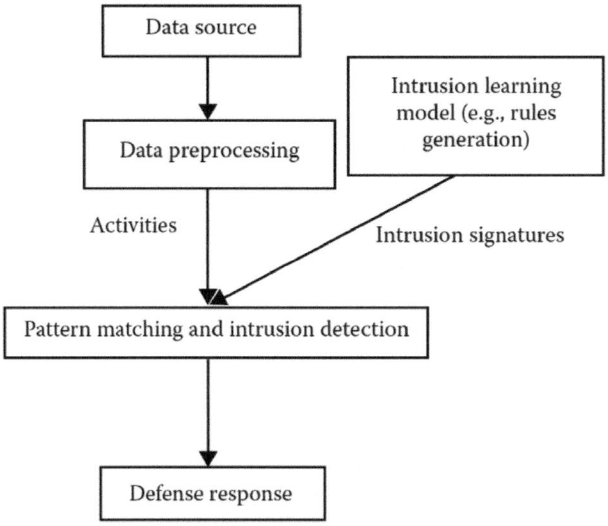

Figure 2.
Sequence of execution of misuse or signature detection modules.

We present a variety of misuse detection techniques that are based on machine learning methods. In the following, we discuss some examples of machine learning methods applied in misuse detection systems.

3.1 Classification using association rules

Agrawal et al. proposed an elegant approach to discover underlying association rules to identify and then establish causal relationships among attributes that may exist in a multidimensional database [1]. Association rules mining identifies the frequent existing patterns in a dataset. This may help, for example, in designing algorithms for a computer antivirus software. A computer antivirus attempts to identify viruses that exhibit some frequently occurring patterns in a transaction dataset. The use of association rules mining and frequently occurring episodes from the computer audit data and exploiting those rules in feature selection had also been described in the literature [2]. Fuzzy association rules were designed for misuse and signature detection on 1998 DARPA intrusion detection dataset [3]. For the purpose of feature selection, 41 features were extracted for each connection record that included 24 different attack types. The attack traffic in the network was essentially of four types: (i) denial of service (DoS), (ii) remote to user (R2L), (iii) user to root and (iv) probes. Including the normal traffic in the network, the association rule mining algorithms extracted the essential features of five types of network data—four categories of attack traffic and one type of normal traffic.

3.2 Artificial neural networks

In a connectionist approach, ANNs carry out the task of pattern recognition and pattern matching using very complex nonlinear transformation functions and the use of multiple hyperplanes separating data of one class from the other. The dynamic nature of the network traffic and the ever-changing characteristics of various attacks on the networks require a very flexible and adaptive misuse detection system that can efficiently and effectively identify a variety of intrusions. Application of ANNs in designing a misuse detection system incorporates the ability to analyze data even if the data may be noisy, distorted, and incomplete. Since ANNs have the ability of learning very accurately from training data, these models can very effectively detect misuse attacks and identify suspicious events in a network. However, this hypothesis is based on the assumption that attackers usually deploy the same approach of an attack on multiple networks, and ANNs can effectively detect similar attacks that had been used by the attackers in the past. Use of ANNs for misuse and signature-based intrusion detection is discussed in [4]. The intrusion detection system (IDS) presented by the author exploits the ability of ANN in classifying nonlinearly separable data into various classes even if the data sources are noisy and limited. The ANN, which is also called a feed-forward multi-layer perceptron (MLP), is equipped with four fully connected layers of nodes with nine nodes in the input layer, two nodes in the output layer, and two hidden layers between the input and the output layers. The two nodes in the output layer are used to indicate the classification results of the network traffic—the normal traffic data being classified with a label of 1, and the attack traffic with a label of 0. Nine important features were extracted by the ANN from the network event data. The network event data are gathered from the data packets transmitted over the network. The nine features of network data traffic that are extracted by the ANN are: (i) protocol, (ii) source port number, (iii) destination port, (iv) source internet protocol (IP) address, (v) destination IP address, (vi) internet control message protocol (ICMP) type, (vii) ICMP code, (viii) data payload length, and (ix) data payload content. Each record of network traffic data was first preprocessed, its features were extracted, and then the features having categorical values were transformed into some standardized numeric values. Around 10,000 network traffic records were synthetically generated for the training and the testing of the ANN model, and approximately 3000 among those records were detected to be anomalous.

3.3 Support vector machines

Because of its intrinsic characteristics, support vector machines (SVMs) are capable of minimizing the structural risk of a dataset by reducing its classification error on unseen records, unlike ANNs which focus more on minimization of empirical risk of the dataset. In order to achieve its goal, a model based on the SVM approach determines its number of parameters based on the margin that separates the data points. This margin is determined by the number of support vectors present in the dataset. The support vectors are those data points that lie nearest to the hyperplane but belong to different classes. In contrast to an ANN, the number of parameters in an SVM model does not depend on the number of feature dimensions in the dataset. This unique property of an SVM makes it so powerful in many practical machine learning applications. In the context of intrusion detection applications, SVMs present two distinct advantages over their ANN counterparts. SVM models execute much faster and they are more scalable. High speed in execution is crucial for detecting attacks in real-time, while scalability is a mandatory requirement for deployment in a complex cyberinfrastructure. Moreover, SVM models can be made to adapt fast based on changes introduced in the training dataset. This feature of SVM is critical when the patterns in the attack traffic change very rapidly. Mukkamala et al. demonstrated how SVM models can be deployed for the purpose of detection of an attack and misuse patterns in context to computer security breaches [5]. The security breaches considered by the authors were bugs in system software bugs, hardware or software failures, incorrect system administration procedures, or failure of the system authentication. For the purpose of building the SVM model, the authors used a training set of 699 data points that contained some records representing actual attack traffic, some records that represented probable attacks, and remaining records exhibiting normal traffic patterns. Eight features were extracted after the initial cleaning and preprocessing phase of the data. Finally, all feature values for each record were normalized to [0, 1]. The test dataset consisted of 250 data points and 8 features. In the confusion matrix yielded by the classification model produced a precision value of 85.53% on the training dataset, and the corresponding value for the test dataset was 94%. This experiment clearly demonstrated the fact that SVM is, in general, more efficient and accurate in identifying misuse and signature-based attack traffic than its ANN counterpart. It also validated the hypothesis that an SVM can effectively simulate security scenarios using its component to adapt to a given information system. Once adapted to a given system, an SVM model can carry out real-time detection of attack traffic, and minimize false alarms while yielding a very high true detection rate.

3.4 Decision tree and classification and regression tree

Decision tree is a nonparametric machine learning method of model building that does not impose any preconditions or requirements on the data. A typical decision tree uses a classification algorithm that labels a data point based on the feature values in the data record corresponding to that node in the decision tree. In order to arrive at a classification decision corresponding to a leaf node in a decision tree, one has to trace the path from the root node to the leaf node. The trace of the path from the root node to a given leaf node can then be converted to a classification rule. If designed optimally, decision trees can yield high classification accuracy, while they involve less complexity in implementation, and have the ability to model intuitive knowledge stored in a high-dimensional dataset. It is precisely these characteristics that make decision trees a very popular choice in many real-world applications. Among the decision tree algorithms, CART represents trees in a form of binary recursive partitioning. It classifies objects or predicts outcomes by selecting from a large number of variables. The most important of these variables determine the outcome variable. Kruegel and Toth proposed a signature and misuse detection system following a decision tree-based approach [6]. In the scheme proposed by the authors, the original rules were partitioned into a smaller subset of rules in such a way that the analysis of a single subset is enough for each input element in the signature detection system [6]. The decision tree algorithm was utilized for detecting the feature that most effectively discriminated against the rule sets of different classes. The algorithm is executed in parallel for evaluating each feature on all the rules in a subset. In the decision tree, the root node corresponded to the universal set of rules. In other words, the root node contained all the rules. The children nodes represented the direct subsets of rules that were partitioned from the rule set based on the first feature in the dataset. The splitting of the nodes in the tree continued till a stage was reached where each node was found to contain one rule only. Labeling was done on each node using the feature that was used for splitting the node. Each directed edge emanating from a node and impinging on its child was marked with the value of the feature specified in the child node. Each leaf node contained either one rule or a set of rules that were not distinguishable by the features in the dataset. During splitting, the sequence of features encountered had an impact on the shape and depth of the tree structure. The authors had also proposed an algorithm that generated a decision tree for detecting malicious events using a limited number of comparisons on the set of rules extracted. Chebrolu et al. used KDD cup 1999 intrusion detection dataset to build a classification and regression tree (CART) [7]. The dataset included 5092 cases and 41 variables. There were 208,772 possible splits in the CART algorithm. Gini index was used for determining the optimal splitting at the nodes.

3.5 Bayesian network classifier

The major shortcoming of most of the rule-based approaches to classification is that these methods treat each event in isolation and never consider the entire gamut of events together taking into account their contextual and temporal relationships. A rule is derived based on the signature of a packet. The signature of a packet is determined using a set of protocols. Many a time, the signature exhibited by a subset of packets belonging to the activities of a malicious user may match that of a normal user; rule-based misuse detection systems often suffer from high rates of false alarm. In the case of a false alarm, the intrusion detection system erroneously identifies an activity in a network as malicious while the activity is actually perfectly normal. Bayesian network (BN)-based models get rid of this problem of rule-based detection systems. Using Bayesian statistics, BN represents problems in networks by specifying the causal relationships between subsets of variables. Typically, a BN is presented as a directed graph that does not contain any cycle. Hence, a BN is also referred to as a DAG-directed acyclic graph. Each node in a BN represents a random variable. A random variable is a variable that can assume a set of values; each value has a specified probability of occurrence. Each arc in a BN depicts a causal relationship with the dependence of the child node on the parent node being expressed as a conditional probability value. The head node and the tail node of an arc are referred to as the parent node and the child node respectively. For example, if in a BN, there is an arc $X_1 \rightarrow X_4$, then X_1 is the

Detection mechanism	Input data format	Detection level	References
Rule-based signature detection	Frequency of system calls, offline	Host	[2]
Fuzzy association rules	Frequency of system calls, online	Host	[3]
Artificial neural networks	TCP/IP packets, offline	Host	[4, 12]
Support vector machines	TCP/IP packets, offline	Network	[5]
Linear genetic programs	TCP/IP packets, offline	Network	[3]
Decision tree	TCP/IP packets, online	Network	[6]
Classification and regression trees	Frequency of system calls, offline	Host	[7]
Statistical method	Executables, offline	Host	[11]
Bayesian networks	Frequency of system calls, offline	Host	[7]

Table 1.
Misuse or signature-based detection schemes.

predecessor of node X_4, and X_4 is the descendent of node X_1. In the example BN, the node X_1 has no predecessor. However, it has three descendant nodes: X_2, X_3, and X_4. Along with BN, the conditional probability table (CPT) presents the dependencies on the net for each variable/node. For each variable/node, the conditional probability P(variable | parent (variable)) is given in CPT for each possible combination of its parents [8–10]. Chebrolu et al. investigated the performance of a feature selection and classification algorithm using BN [7]. Markov blanket method was used to find the most significant feature set in a training dataset that included five classes of network traffic: normal, probe, DOS, U2R, and R2L.

3.6 Naïve Bayes

The naïve Bayes classifier makes the assumption of class conditional independence. Given a data sample, its features are assumed to be conditionally independent of each other. This is in contrast with a BN that assumes dependencies among the features. Schultz et al. used the naïve Bayes approach to detect new, previously unseen malicious executables accurately and automatically [11].

Most of the machine learning methods for misuse and signature detection are in the initial stages of research and are yet to find any commercial deployment. Moreover, feature selection before traffic classification is a challenging task. Detection quality heavily depends on the experience and knowledge of the security experts dealing with the problem. It also depends on an exhaustive testing and refining process. The use of decision trees for the selection of a significant feature subset has only partially solved this problem. **Table 1** summarizes the signature-based detection schemes we discussed in this section. We have categorized the schemes based on their approach, input data used, and level of detection.

4. Anomaly detection

When a novel attack is launched on a network, misuse detection systems cannot detect the attack as the attack signature is not present in the existing database of attack signatures. However, an anomaly detection system has the ability to detect new and unseen attacks and raise an early alarm before any substantial damage to the network could be done by the attack. Like the misuse detection approach, anomaly detection relies on determining a clear boundary between the normal and the anomalous traffic. The profile of the normal behavior is assumed to be significantly different from that of the anomalous behavior. The profile of the normal events and the normal traffic should preferably satisfy a set of criteria in the sense that it must contain a very clearly defined normal behavior. For example, the normal behavior specification must include the IP address or the hostname of a computing machine, or it should include a virtual local area network's (VLAN) details to which it belongs and have the ability to track the normal behavior of the target environment sensitively. In addition, the normal profile should include the following details: (i) occurrence patterns of some specific system calls in the application layer of the communication protocol stack, (ii) association of data payload with different fields of application protocols, (iii) connectivity patterns between secure servers and the Internet and (iv) the rate and the burst length distributions of all traffic types [13]. In addition, profiles based on a network must be adaptive and have the ability of self-learning from complex and challenging network traffic to preserve the accuracy and in achieving a low false acceptance rate (FAR).

In a large data network, detection of malicious and anomalous traffic is a complex task that poses some significant critical challenges. It is difficult to analyze and monitor a huge volume of traffic that contains network data with a very high-dimensional feature space. Such monitoring and analysis of network traffic data calls for highly efficient computational algorithms in data processing and pattern learning. Moreover, the anomalous traffic in a network exhibits a common behavior. In large volume network traffic data, the malicious and anomalous traffic of the same type tends to occur repeatedly, while the number of occurrences of malicious and anomalous data is much smaller than the number of occurrences of normal data. This makes the network traffic data highly imbalanced. It is also difficult, if not impossible, to determine accurately a normal region, or define the boundary between the normal and the anomalous traffic. To complicate the issue further, the concept of anomaly varies among different application domains. In many situations, labeled anomalous data are not available for the training and validation processes. Training and testing data contain the noise of unknown distributions, and the normal and anomalous behavior constantly changes. All these issues make anomaly detection in a network a particularly difficult task.

5. Machine learning in anomaly detection

Figure 3 depicts the schematic diagram of a typical anomaly detection system. Anomaly detection systems broadly work in the five steps: (i) data collection, (ii) data preprocessing, (iii) normal behavior learning phase, (iv) identification of misbehaviors using dissimilarity detection techniques and (v) security responses. In a large-scale network, the data collection phase involves a large volume of data to be collected from the network. In the data preprocessing phase, the volume of data is reduced as this step includes feature selection, feature extraction, and finally dimensionality reduction processes.

Machine learning algorithms can be very effective in building normal profiles and then in designing intrusion detection systems based on anomaly detection approach. In the anomaly detection approach, the network traffic data belonging to a normal class are usually available for training the model. However, in most of the applications, labeled data for anomalous traffic are not available. We have already seen that supervised machine learning algorithms need attack-free training data. In other words, supervised learning needs labeled network data for both types of traffic—normal and attack. However, in most of the real-world situations, such

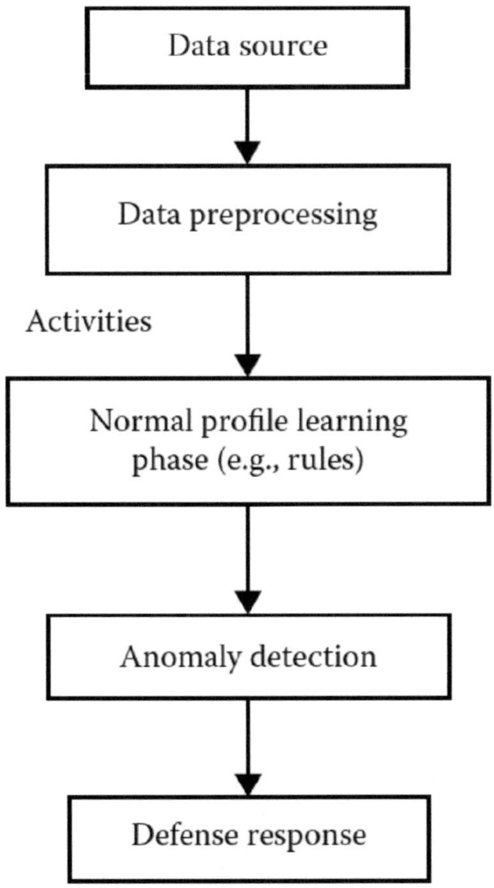

Figure 3.
Sequence of execution of modules in an anomaly detection system.

prelabeled training data for both classes are very difficult to get. In most cases, not only are the prelabeled training data not available, but also the traffic data in networks exhibit highly imbalanced characteristics. A large majority of normal traffic record is mixed with a tiny minority of attack traffic records. To make the challenge even bigger, with the change in the network environment, patterns of normal traffic also exhibit substantial changes. The significant difference in the characteristics of training and test datasets most often leads to high false positive rates (FPRs) for supervised intrusion detection systems (IDSs). Unsupervised learning methods as adopted by anomaly detection systems can potentially get rid of this problem by building a normal profile of network traffic and by defining a normal state of the system. Any deviation from the normal state indicates the presence of an anomalous activity in a network. Hence, semi-supervised and unsupervised machine learning methods are frequently deployed in real-world security applications [14].

5.1 Rule-based anomaly detection

In misuse detection, rules depict the strength of correlation between the conditions of the attributes and class labels. In the context of anomaly detection, the rules are the descriptors of normal profiles of users, application and system programs, and other resources in the computing and network infrastructures. An anomaly detection system is expected to raise an alarm of a potential attack if it observes any inconsistency among the current activities of the programs and the users with the established rules in the system. For an anomaly detection system to work effectively, it is critical to have an exhaustive set of rules working. The use of associative classification and association rules in anomaly-based intrusion detection systems is quite common. A number of propositions exist in the literature that has exploited the power of association rules in designing anomaly detection models [2, 15, 16]. Anomaly detection systems using association rules broadly work in two steps. In the first step, effective data mining operations are carried out on the system and network audit data for identifying consistent and useful patterns of the behaviors of the programs and the users. In the second step, robust classifiers are inductively learned using the training dataset on the relevant features in the patterns to recognize any anomalous behavior in the system or in the network traffic. The concept of frequent episodes is presented in [17]. Lee and Stolfo utilized the concept of frequent episodes introduced in [17] to characterize the audit sequences occurring in normal data [2]. Based on the frequent episodes in the network, the authors designed a small set of rules that could effectively capture the frequent behaviors in those sequences. During the monitoring phase of the detection system, the event sequences that were found to violate the rules are identified as the anomalous events in the cyberinfrastructure.

5.2 Fuzzy rule-based anomaly detection

The anomaly detection systems working on the association rules use a deterministic value or an interval to quantify the rules. In such a scenario, the normal and anomalous records are separated by clearly defined and sharp boundaries in the n-dimensional feature space, where n is the number of features in the dataset. However, such a crisp separation poses a significant challenge in correctly detecting the normal audit records in situations where these normal data deviate from the established association rules by a small margin. This problem is handled by introducing fuzzy logic in designing the association rules, and thereby incorporating flexibility in the operations of rule-based anomaly detection systems. Moreover, many of the features may be ordinal or categorical in nature, thereby making the design of association rules based on crisp and deterministic values of the features a well-neigh impossible proposition. Hence, the introduction of fuzziness in the association rules becomes mandatory. For example, a rule may contain the connection duration of a user's process by using the following expression, such as "connection duration = 3 min" or "1 min ≤ connection duration ≤ 4 min." Luo and Bridges investigated the fuzzy rule-based anomaly detection using real-world data and simulated dataset [18]. The real-network traffic data were collected by the Department of Computer Science at Mississippi State University by *tcpdump* [19]. Four features were extracted from the data. These features were denoted as: SN, FN, RN, and PN. SN, FN, and RN denote, respectively, the number of SYN, FIN, and FST flags appearing in the TCP packet headers in the last 2 seconds. PN denotes the number of destination ports in the last 2 seconds. Three fuzzy sets were designed, which were given names: LOW, MEDIUM, and HIGH. Each feature was divided into these three fuzzy sets. Fuzzy association rules were derived from the dataset based on the first three features of the data, and fuzzy frequency episode rules were designed for the last feature. Network traffic data in the afternoon of a given day were used in training of the model and in deriving the fuzzy rules in the normal traffic data. The traffic data from the afternoon, evening, and night on the same day were used for testing and anomaly detection. For testing the model, a similarity function was used to compare the normal patterns with the anomalous patterns.

5.3 Artificial neural networks

Artificial neural networks (ANNs) allow for generalization in incomplete data and enable the detection of anomalous behavior in anomaly detection systems. The standard feed-forward multi-layer perceptron (MLP) with the ability of backpropagation of errors is particularly suited for carrying out anomaly detection. In the forward propagation phase, the ANN is trained on the training dataset. The data are fed into the network through the nodes in the input layer. The nodes at each layer are activated and their output passed on to the nodes in the next layer till the output values come out of the nodes at the output layer. The output values produced by the output layer nodes are then compared with the desired or target values at the corresponding nodes. The difference between the actual output value and the target output value signifies the error at the node at the output layer. The error values are backpropagated through the links in the network from the nodes at the output layer back to the nodes at the input layer so that the weights in the links and the biases at the nodes can be updated. This process of forward and back propagation continues until the error values at the output nodes fall below a threshold value. At this point the training process completes. Ghosh et al. [12, 20] and Liu et al. [21] applied ANNs in anomaly detection methods in computer networks.

5.4 Support vector machines

Support vector machines (SVMs) outperform ANNs in many situations as they have the ability to attain the global optimum state more efficiently and can control the model overfitting problem more effectively by fine-tuning the model parameters. SVMs can be gainfully deployed in anomaly detection by training them on datasets containing attack traffic and normal traffic. This is a supervised way of learning for SVMs. However, SVMs can also be applied effectively in an unsupervised way of identifying anomalous traffic in a network. Chen et al. used BSM audit data from the 1998 DARPA intrusion detection evaluation datasets and trained an SVM-based anomaly detection system using the dataset [22]. Hu et al. presented a comparative study on the performance of a robust support vector machine (RSVM) and a conventional SVM based on the nearest neighbor classification in separating normal traffic from attack traffic generated by various computer programs [23]. The results presented by the authors clearly showed that RSVMs had higher detection accuracy with a much lower value of false positives as compared with their conventional SVM counterparts. RSVMs also exhibited higher generalization ability in extracting information from noisy data.

5.5 Nearest neighbor-based learning

Nearest neighbor-based machine learning programs assume that the normal pattern of an activity displays a close displacement measured by a distance metric, while anomaly data points lie far from this neighborhood. K-nearest neighbor (KNN) method is a classification approach that uses a voting score among all the neighbors of a given data point in determining its class membership. The KNN learning-based anomaly detection method is effective only if the value of k is more than the frequency of occurrence of any anomalous data in traffic audit dataset, and the Euclidean distance between the anomalous data groups from the normal data group is large in the n-dimensional feature space of the traffic dataset. In the literature, several anomaly detection approaches have been proposed using different variants of the basic nearest neighborhood-based classification method. These methods use different definitions of the nearest neighbor for the purpose of detection of anomalous traffic. Liao and Vemuri presented a KNN classifier model to classify the behavior of computer programs into two types—normal and anomalous [24]. In the proposed scheme, the behavior of a program was represented by the number of system calls made by the program. While every system call was treated as a word, the set of all system calls made by a program over its entire life span of execution was compiled as a document. The programs were subsequently classified into normal or anomalous classes using a KNN classifier constructed using document classification methods on the documents. The experiments were performed using the BSM audit data in the 1998 DARPA intrusion detection evaluation datasets. In the training phase, 3556 normal programs and 49 distinct system calls in 1 simulation day were used. The test audit data were scanned for programs to measure the distance. The distances were then sorted in the increasing order of their magnitudes and the top k scores were selected for the k nearest neighbors for each of the records in the test audit data. For the purpose of anomaly detection, a threshold value of the average of the top k distances for each record in the test dataset was determined. In their experiments, authors tried out different values of the threshold distance and the k values so as to determine the most optimal performance of the KNN classifier as depicted by its receiver operating characteristics (ROC) curve. The KNN algorithms were found to detect 100% of the attacks while keeping a *false positive rate* (FPR) at a very low value of 0.082% with $k = 5$ and a threshold value of 0.74.

5.6 Hidden Markov model and Kalman filter

Hidden Markov model (HMM) considers transition properties of events. In network security applications it can be effectively deployed for detecting anomalous activities and events. In anomaly detection, HMMs can very accurately model the temporal variations in program behavior [25–27]. Before the deployment of an HMM in anomaly detection, the definition of a normal sate of activity S and a dataset of normal observable events O are to be decided upon. Starting from the initial state of S, and given a sequence of observations Y, the HMM searches for a sequence X that contains all normal states, and that has a predicted observation sequence that is most similar to Y with a computed probability value. If this computed probability value is smaller than a predefined threshold value, the sequence Y is assumed to have led the system to an anomalous state. Warrender et al. proposed an HMM-based anomaly detection model using publicly available datasets on systems calls from nine programs [25]. The datasets used were MIT LPR and UNM LPR [25]. An HMM with 40 states was designed. These 40 states represented 40 system calls that were present in all those nine programs. The HMM was designed in a fully connected manner so that transitions were possible from any given state to any other state in the model. The Baum-Welch algorithm was applied to fine-tune the parameters of the HMM using the training dataset [28]. The Baum-Welch algorithm works on the principles of dynamic programming and it is a variant of expectation maximization (EM) algorithm. The Viterbi algorithm was utilized to find out which choice of states maximizes the joint probability distribution given the trained parameter matrices of the HMM [29]. In other words, the Viterbi algorithm identifies the most likely state, given a dataset and a trained HMM model. The authors contend that for a well-designed HMM, a sequence of system calls that represents normal activities will lead to state transitions and output values that are highly likely; on the other hand, a sequence of system calls that represents an anomalous activity will lead to state transitions and output values that are unusual. Hence, in order to detect anomalous events in a network, it is sufficient to track unusual state transitions and abnormal output values. The experimental results indicated that the

HMM could detect anomalous traffic efficiently and effectively with a low value of mismatch rate. In general, training of an HMM is a very time-consuming process as it requires multiple epochs (i.e., passes) through the records in a training dataset. Since all the transition probabilities corresponding to long sequences of state transitions are needed to be stored, training an HMM is a memory-intensive operation as well. Soule et al. presented an anomaly detection method in a large-scale data network [30]. The detection scheme analyzed the traffic patterns in a network, and computed the state the network using a Kalman filter. A Kalman filter is a set of mathematical equations that implements a predictor-corrector type estimation that is optimal [31]. The optimality here refers minimization of error covariance. The Kalman filter used in the anomaly detection filtered out the normal traffic state by comparing the predictions made by the current traffic state to an inference of the actual traffic state. The residual process is then analyzed for possible anomalies.

5.7 Clustering-based anomaly detection

Supervised learning methods for the detection of anomalous activities in a network require prior labeling of the traffic types. However, it is very difficult to have prior labeling of audit data in real-world network environments. Signature-based detection suffers from this problem as carrying out a manual classification in a huge volume of network traffic to identify a small number of attack traffic records poses a significant challenge. Unsupervised learning-based anomaly detection methods do suffer from this drawback as these methods can work on unlabeled network traffic data. These methods attempt to detect malicious traffic in a network even without any prior knowledge about the traffic data labels. Unsupervised learning-based anomaly detection methods work under the following premise: in a network, characteristics of traffic are highly imbalanced—normal traffic constitutes a vast majority, while anomalous traffic represents a tiny minority. Moreover, attack traffic and the normal traffic exhibit similar statistical distributions in their respective group, while the distributions of the two groups are different from each other. Learning from an imbalanced data so that the anomalous and normal traffic can be categorized into two different clusters is the prime focus in unsupervised anomaly detection methods. Hence, cluster-based anomaly and outlier detection is the most fundamental approach in an unsupervised intrusion detection method. Portnoy et al. proposed a clustering-based anomaly detection method using DARPA knowledge discovery in databases (KDD) Cup 1999 dataset [32]. DARPA KDD Cup 1999 dataset consisted of a network traffic record of 4,900,000 data points. The dataset contained 25 different types of traffic—24 attack types and 1 normal traffic. Each data point represented a set of extracted feature values from a connection record obtained between different IP addresses of hosts during a period of time in which attacks were simulated in a network. The authors observed that clustering with unlabeled data resulted in a lower detection rate of attacks than attack classification using a supervised learning method. However, unsupervised detection methods on unlabeled data can potentially detect unknown attacks through an automated or semi-automated process that cannot be done using supervised detection methods.

5.8 Random forests

Random forests are powerful machine learning models based on ensemble approach. They build multiple decision trees by randomly choosing a subset of features and then combine those decision tree results to arrive at a much more robust prediction. Due to their higher accuracy of prediction, random forests have been deployed in a variety of applications including multimedia information retrieval,

Detection mechanism	Input data format	Detection level	References
Statistical methods	Frequency of system calls, offline	Host	[36, 37]
Statistical methods	TCP/IP packets, online	Network	[30, 38, 40]
Clustering algorithms	Frequency of system calls, online	Network	[25, 32, 33]
Information theoretic	TCP/IP packets, offline	Network	[37, 42, 45]
Association rules	TCP/IP packets, offline	Host	[2, 15–17]
Fuzzy association rules			[18]
Kalman filter	TCP/IP packets, online	Network	[30]
Hidden Markov model	Frequency of system calls, offline	Host	[25–27]
Artificial neural network	Executables, offline	Host	[12, 20, 21]
Principal component analysis	Frequency of system calls, offline	Network	[42–44]
Support vector machine	TCP/IP packets, offline	Network	[22, 23]
K-nearest neighbors	Frequency of system calls, offline	Host	[24]
Random forests	TCP/IP packets, offline	Network	[33, 34]

Table 2.
Anomaly-based detection schemes.

network security and intrusion detection systems design. The algorithms used in random forests usually yield higher accuracy, and they work very efficiently on large datasets with high-dimensional feature space. Traffic in a large network is an example of a large volume of high-dimensional data, and such data can be very effectively classified in real-time by random forest-based classification approach. The use of random forest algorithms for detecting outliers in datasets containing network traffic without attack-free training data has been proposed in the literature [33, 34].

5.9 Other machine learning methods in anomaly detection

Other machine learning methods have been proposed for learning the probability distribution of data and in applying statistical tests to detect outliers. Eskin proposed a mixture probability model on normal and anomalous data based on expectation maximization (EM) algorithms [35]. Other statistical machine learning methods have been investigated in anomaly detection applications, such as mean and variance [4, 30], Hotelling's T^2 test and the Chi-square test [36, 37], Hellinger score [38], histogram density [39], Bayesian law [40], cumulative summation (CUMSUM) and statistical test [30]. Ye et al. used a series of probability techniques of anomaly detection, including decision tree, Hotelling's T^2 test, Chi-square multivariate test, and Markov chain in an information system for detecting intrusions [41]. Network-wide anomaly detection using principal component analysis (PCA) has been proved very effective [42–44]. Several studies have also found that a wide range of anomalies in networks can be detected by computing the entropy in the network flow and feature distributions [37, 42, 45]. **Table 2** presents a summary of the anomaly-based detection schemes.

6. Machine learning in hybrid detection

Since misuse detection systems work on matching already known attack signatures with the current events in a network, they usually have high detection rates and low false alarm rates. However, these systems cannot detect novel attacks. On the other hand, anomaly detection systems define normal sates in a network and then detect system states that significantly differ from the normal states. Any state that significantly differs from the normal state of the network indicates the possible event of an attack. The anomaly detection system can detect new attacks launched on a network. There is a challenge in the anomaly detection system design. If the normal state patterns do not significantly differ from patterns exhibited by any anomalous state, the attack state will go undetected. This leads to an increase in the false alarm rate. Hence, it is critical to design a normal state in such a way that while the detection rate is maximized, the number of false alarms does not exceed beyond an acceptable limit. If the normal state is too wide, then the detection rate will suffer. On the other hand, too narrow a normal state will lead to a high false alarm rate. The hybrid detection approach combines the adaptability and the powerful detection ability of an anomaly detection system with the higher accuracy and reliability of the misuse detection approach.

Designing an efficient and accurate hybrid detection system involves two critical issues: (i) the most ideal misuse or anomaly detection systems are to be first identified that can be integrated with anomaly detection systems, so that hybrid detection is possible and (ii) the two systems are to be integrated in the most optimal way so that the balance between the detection rate and false alarm rate is achieved while retaining the ability of detecting novel attacks.

The selection of misuse and anomaly detection systems for designing a hybrid detection system is dependent on the application in which the detection system is to be deployed. Following a combinational approach, the integration of an anomaly detection system with a misuse detection counterpart has been classified into four categories [46, 47]. These types are: (i) anomaly-misuse sequence detection, (ii) misuse-anomaly sequence detection, (iii) parallel detection and (iv) complex mixture detection (**Figure 4**). The complex mixture model is highly application-specific.

Barbara et al. presented a hybrid detection system on the principle of the anomaly-misuse sequence [48]. The proposed system, which is known as audit data analysis and mining (ADAM), minimizes false alarm rates by not raising any alarm for those patterns that are not classified attacks by the misuse detection system.

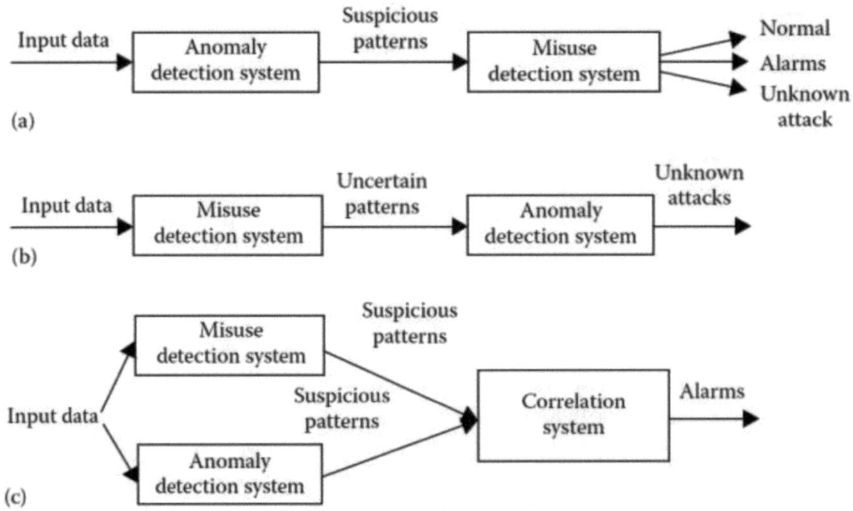

Figure 4.
Three categories of hybrid detection systems. (a) Anomaly-misuse sequence, (b) misuse-anomaly sequence, (c) parallel detection system (adapted from [43]).

Detection mechanism	Input data format	Detection level	References
Random forests	TCP/IP packets, online	Network	[46, 47, 49]
Association rules	TCP/IP packets, online	Network	[48]
Association rules	Frequency of system calls, online	Host	[2]
Cooperating agents	TCP/IP packets, online	Network	[52–56]
Correlation	TCP/IP packets, online	Network	[50]
Clustering	TCP/IP packets, offline	Network	[51]
Statistical analysis and ANN	Sequences of system calls, offline	Host	[55]
ANN	TCP/IP packets, online	Network	[12]

Table 3.
Hybrid intrusion schemes based on machine learning.

Misuse-anomaly sequence detection systems primarily focus on detecting novel attacks that are missed by the misuse detection module. The machine learning algorithms used by these hybrid detection models are mainly based on different variants of random forests [47, 49]. Anderson et al. proposed the design of a parallel intrusion detection system that provided a very accurate and robust detection decision by correlating the outputs of the misuse detection and the anomaly detection modules [50]. Agrawal et al. proposed an illustrative complex intrusion detection system [51]. The system worked on the AdaBoost algorithm of classification and both the misuse and the anomaly detection systems are trained on the training data simultaneously. The detection results on the test data are also presented separately for the misuse detection module and the anomaly detection module. Sen et al. proposed various architecture of complex detection systems based on cooperating agents [52–54]. The audit trails in the basic security module (BSM) of a Solaris system were exploited by Endler in designing a hybrid detection system [55]. An ANN-based hybrid detection system for detecting both signature-based and anomaly-based attacks is proposed by Ghosh and Schwartzbard [12]. Lee et al. presented a data mining-based hybrid intrusion detection system for identifying attack traffic from the audit data in a host [2]. **Table 3** presents a summary of the hybrid detection systems discussed in this section.

7. Conclusion

In this chapter, we have discussed various approaches to misuse and anomaly detection systems design using machine learning and data mining techniques. Some of the well-known systems in the literature have also been reviewed briefly. We have also discussed the pros and cons of various systems in context to their applications and deployment in real-world networks.

A fundamental challenge in designing an intrusion detection system is the limited availability of appropriate data for model building and testing. Generating data for intrusion detection is an extremely painstaking and complex task that mandates the generation of normal system data as well as anomalous and attack data. If a real-world network environment, generating normal traffic data is not a problem. However, the data may too privacy-sensitive to be made available for public research.

Classification-based methods require training data to be well balanced with normal traffic data and attack traffic data. Although it is desirable to have a good mix of a large variety of attack traffic data (including some novel attacks), it may not be feasible in practice. Moreover, the labeling of data is mandatory with attack and normal traffic data clearly distinguished by their respective labels.

Unlike classification-based approaches, which are mostly used in misuse detection, unsupervised anomaly detection-based approaches do not require any prior labeling of the training data. In most of the cases, the attack traffic constitutes the sparse class, and hence, the smaller clusters are most likely to correspond to the attack traffic data. Although unsupervised anomaly detection is a very interesting approach, the results produced by this method are unacceptably low in terms of their detection accuracies.

In a pure anomaly detection approach, the training data are assumed to be consisting of only normal traffic. By training the detection model only on the normal traffic data, the detection accuracy of the system can be significantly improved. Anomalous states are indicated by only a significant state change from the normal sate of the system.

In a real-world network that is connected to the Internet, an assumption of attack-free traffic is utopian. A pure anomaly detection system can still be trained on training data that include attack traffic. In that case, those attack traffic data will be considered as normal traffic and the detection system will not raise an alert when such traffic is encountered in real-world operations. Hence, in order to increase the detection accuracy, attack traffic should be removed from the training data as much as possible. The removal of attack traffic from the training data can be done using updated misuse detection systems or by deploying multiple anomaly detection systems and combining their results by a voting mechanism.

For an intrusion detection system that is deployed in a real-world network, it is mandatory to have a real-time detection capability under a high-speed, high-volume data environment. However, most of the cluster techniques used in unsupervised detection require quadratic time. This renders their deployment infeasible in practical applications. Moreover, the cluster algorithms are not scalable, and they need the entire training data to reside in the memory during the training process. This requirement puts a restriction on the model size. The future direction of research may include studies on the scalability and performance of anomaly detection algorithms in conjunction with the detection rate and false positive rate. Most of the currently existing propositions on intrusion detection have not paid adequate attention to these critical issues.

Author details

Jaydip Sen* and Sidra Mehtab
School of Computing and Analytics, NSHM Knowledge Campus, Kolkata, India

*Address all correspondence to: jaydip.sen@acm.org

References

[1] Agrawal R, Imielinski T, Swami A. Mining association rules between sets of items in large databases. In: Proceedings of the ACM SIGMOD International Conference on Management of Data. Washington, DC: ACM; 1993. pp. 207-216

[2] Lee WK, Stolfo SJ, Mok KW. A data mining framework for building intrusion detection models. In: Proceedings of the IEEE Symposium on Security and Privacy. Oakland, CA: IEEE; 14 May 1999. pp. 120-132. DOI: 10.1109/SECPRI.1999.766909

[3] Abraham A, Grosan C, Martin-Vide C. Evolutionary design of intrusion detection programs. International Journal of Network Security. 2007;**4**(3):328-339. DOI: 10.6633/IJNS.200705.4(3).12

[4] Cannady J. Artificial neural networks for misuse detection. In: Proceedings of the National Information Systems Security Conference (NISSC'98). Washington, DC; 6-9 October 1998. pp. 441-454

[5] Mukkamala S, Janoski G, Sung AH. Intrusion detection using neural networks and support vector machines. In: Proceedings of the International Joint Conference on Neural Networks (IJCNN'02). Honolulu, HI; 12-17 May 2002. pp. 1702-1707. DOI: 10.1109/IJCNN.2002.1007774

[6] Kruegel C, Toth T. Using detection trees to improve signature-based intrusion detection. In: Proceedings of the 6th International Workshop on Recent Advances in Intrusion Detection. Pittsburgh, PA; 8-10 September 2003. pp. 173-191. DOI: 10.1007/978-3-540-45248-5_10

[7] Chebrolu S, Abraham A, Thomas JP. Feature deduction of intrusion detection systems. Computers & Security. 2005;**24**:295-307. DOI: 10.1016/j.cose.2004.09.008

[8] Cooper GF, Herskovits E. A Bayesian method for the induction of probabilistic networks from data. Machine Learning. 1992;**9**:309-347. DOI: 10.1007/BF00994110

[9] Verma T, Pearl J. An algorithm for deciding if a set of observed independencies has a causal explanation. In: Proceedings of the 8th International Conference on Uncertainty in Artificial Intelligence. Stanford, CA; July 1992. pp. 323-330. DOI: 10.1016/B978-1-4832-8287-9.50049-9

[10] Pearl J, Wermuth N. When can association graphs admit a causal interpretation? In: Proceedings of the 4th International Workshop on Artificial Intelligence and Statistics. Fort Lauderdale, FL; 1993. pp. 141-150. DOI: 10.1007/978-1-4612-2660-4_21

[11] Schultz MG, Eskin E, Zadok E, Stolfo SJ. Data mining methods for detection of new malicious executables. In: Proceedings of IEEE Symposium on Security and Privacy (S&P'01). Oakland, CA. Anaheim, CA; 14-16 May 2000. DOI: 10.1109/SECPRI.2001.924286

[12] Ghosh AK, Schwartzbard A, Schatz M. Learning program behavior profiles for intrusion detection. In: Proceedings of the 1st USENIX Workshop on Intrusion Detection and Network Monitoring. Santa Clara, CL; 9-12 April 1999. pp. 51-62

[13] Gong F. Deciphering Detection Techniques: Part II. Anomaly-Based Intrusion Detection. Santa Clara, CA, USA: White paper, Mcafee Network Security Technologies Group; 2003

[14] Eskin E, Arnold A, Prerau M, Portnoy L, Stolfo S. A geometric framework for unsupervised anomaly detection: Detecting intrusions in

unlabeled data. In: Jajodia S, Barbara S, editors. Applications of Data Mining and Computer Security. Dordrecht: Kluwer; 2002. pp. 77-101. DOI: 10.7916/D8D50TQT

[15] Lee W, Stolfo SJ. Data mining approaches for intrusion detection. In: Proceedings of the 7th USENIX Security Symposium. San Antonio, TX; 26-29 January 1998. DOI: 10.7916/D86D60P8

[16] Apiletti D, Baralis E, Cerquitelli T, D'Elia V. Characterizing network traffic by means of the NetMine framework. Computer Networks. 2009;**53**(6):774-789. DOI: 10.1016/j.comnet.2008.12.011

[17] Mannila H, Toivonen H. Discovering generalized episodes using minimal occurrences. In: Proceedings of the 2nd International Conference on Knowledge Discovery in Databases and Data Mining. Portland, OR: P. ACM; August 1996. pp. 146-151

[18] Luo J, Bridges SM. Mining fuzzy association rules and fuzzy frequency episodes for intrusion detection. International Journal of Intelligent Systems. 2000;**15**(8):687-703

[19] tcpdump website. Available from: https://www.tcpdump.org

[20] Ghosh AK, Wanken J, Charron F. Detecting anomalous and unknown intrusions against programs. In: Proceedings of the 14th Annual Computer Security Applications Conference (ACSAC'98). Phoenix, AZ; 7-1 December 1998. DOI: 10.1109/CSAC.1998.738646

[21] Liu Z, Florez G, Bridges SM. A comparison of input representations in neural networks: A case study in intrusion detection. In: Proceedings of the International Joint Conference on Neural Networks (IJCNN'02). Honolulu, HI; 12-17 May 2002. DOI: 10.1109/IJCNN.2002.1007775

[22] Chen WH, Hsu SH, Shen HP. Application of SVM and ANN for intrusion detection. Computers and Operations Research. 2005;**32**(10):2617-2634. DOI: 10.1016/j.cor.2004.03.019

[23] Hu WJ, Liao YH, Vemuri VR. Robust support vector machines for anomaly detection in computer security. In: Proceedings of the International Conference on Machine Learning (ICMLA'03). Los Angeles, CL: CSREA; 23-24 June 2003. pp. 161-167

[24] Liao YH, Vemuri VR. Use of *k*-nearest neighbor classifier for intrusion detection. Computers & Security. 2002;**21**(5):439-448. DOI: 10.1016/S0167-4048(02)00514-X

[25] Warrender C, Forrest S, Pearlmutter B. Detecting intrusions using system calls: Alternative data models. In: Proceedings of IEEE Symposium on Security and Privacy. Oakland, CA: IEEE; 10-14 May 1999. pp. 133-145. DOI: 10.1109/SECPRI.1999.766910

[26] Qiao Y, Xin XW, Bin Y, Ge S. Anomaly intrusion detection method based on HMM. Electronics Letters. 2002;**38**(13):663-664. DOI: 10.1049/el:20020467

[27] Wang W, Guan X, Zhang X, Yang L. Profiling program behavior for anomaly intrusion detection based on the transition and frequency property of computer audit data. Computers & Security. 2006;**25**(7):539-550. DOI: 10.1016/j.cose.2006.05.005

[28] Sammut C, Webb GI, editors. Encyclopedia of Machine Learning. Boston, MA: Springer; 2011. DOI: 10.1007/978-0-387-30164-8

[29] Li SA, Jain A, editors. Encyclopedia of Biometrics. Boston, MA: Springer; 2009. DOI: 10.1007/978-0-387-73003-5_592

[30] Soule K, Salamatian K, Taft N. Combining filtering and statistical methods for anomaly detection.

In: Proceedings of the 5th ACM SIGCOMM Conference on Internet Measurement. Berkeley, CA: ACM; 19-21 October 2005. pp. 331-344. DOI: 10.1145/1330107.1330147

[31] Musoff H, Zarchan P. Fundamentals of Kalman Filtering: A Practical Approach. 2nd ed. Reston, VA, USA: AIAA Press. DOI: 10.2514/4.866777

[32] Portnoy L, Eskin E, Stolfo S. Intrusion detection with unlabeled data using clustering. In: Proceedings of ACM CSS Workshop on Data Mining Applied to Security (DMSA). Philadelphia, PA: ACM; November 2001. pp. 5-8. DOI: 10.7916/D8MP5904

[33] Zhang J, Zulkernine M. Anomaly-based network intrusion detection with unsupervised outlier detection. In: IEEE International Conference on Communications. Istanbul, Turkey: IEEE; 11-15 June 2006. pp. 2388-2393. DOI: 10.1109/ICC.2006.255127

[34] Zhang J, Zulkernine M, Haque A. Random forest-based network intrusion detection systems. IEEE Transactions on Systems, Man, and Cybernetics—Part C: Applications and Reviews. 2008;**38**(5):649-659. DOI: 10.1109/TSMCC.2008.923876

[35] Eskin E. Anomaly detection over noisy data using learned probability distribution. In: Proceedings of the 17th International Conference on Machine Learning (ICML'00). Stanford, CA: ACM; 29 June-2 July 2000. pp. 255-262. DOI: 10.7916/D8C53SKF

[36] Ye N, Li X, Chen Q, Emran SM, Xu M. Probabilistic techniques for intrusion detection based on computer audit data. IEEE Transactions on Systems, Man, and Cybernetics - Part A: Systems and Humans. 2001;**31**(4): 266-274. DOI: 10.1109/3468.935043

[37] Feinstein L, Schnackenberg D, Balupari R. Kindred, D. Statistical approaches to DDoS attack detection and response. In: Proceedings of DARPA Information Survivability Conference and Exposition. Washington, DC: IEEE; April 2003. pp. 303-314. DOI: 10.1109/DISCEX.2003.1194894

[38] Yamanishi K, Takeuchi JI. Discovering outlier filtering rules from unlabeled data: Combining a supervised learner with an unsupervised learner. In: Proceedings of the 7th ACM SIGKDD International Conference on Knowledge Discovery and Data Mining. Edmonton, Canada; January 2001. pp. 389-394. DOI: 10.1145/502512.502570

[39] Yamanishi K, Takeuchi J, Williams G, Milne P. On-line unsupervised outlier detection using finite mixtures with discounting learning algorithms. Data Mining and Knowledge Discovery. 2004;**8**(3): 275-300. DOI: 10.1023/B:DAMI.000002 3676.72185.7c

[40] Mahoney MV, Chan PK. Learning nonstationary models of normal network traffic for detecting novel attacks. In: Proceedings of the 8th ACM SIGKDD International Conference on Knowledge Discovery and Data Mining. Edmonton, Alberta, Canada: ACM; 23-26 July 2002. pp. 376-386. DOI: 10.1145/775047.775102

[41] Ye N, Emran SM, Chen Q, Vibert S. Multivariate statistical analysis of audit trails for host-based intrusion detection. IEEE Transactions on Computers. 2002;**51**(7):810-820. DOI: 10.1109/TC.2002.1017701

[42] Lakhina A, Crovella M, Diot C. Mining anomalies using traffic features distributions. Computer Communication Review. 2005;**35**(4): 217-228. DOI: 10.1145/1090191.1080118

[43] Lakhina A, Crovella M, Diot C. Diagnosing network-wide traffic anomalies. In: Proceedings of the 2004 International Conference on Applications, Technologies, Architectures, and Protocols for Computer Communications

(SIGCOMM'04). 2004. pp. 219-230. DOI: 10.1145/1015467.1015492

[44] Ringberg H, Soule A, Rexford J, Diot C. Sensitivity of PCA for traffic anomaly detection. Performance Evaluation Review. 2007;**35**(1):109-120. DOI: 10.1145/1269899.1254895

[45] Lee W, Xiang D. Information-theoretic measures for anomaly detection. In: Proceedings of 2001 IEEE Symposium on Security and Privacy. Oakland, CA; 14-16 May 2000. DOI: 10.1109/SECPRI.2001.924294

[46] Zhang J, Zulkernine M. Anomaly based network intrusion detection with unsupervised outlier detection. In: Proceedings of the IEEE International Conference on Communications (ICC'06). Istanbul, Turkey; 11-15 June 2006. DOI: 10.1109/ICC.2006.255127

[47] Zhang J, Zulkernine M, Haque A. Random-forest-based network intrusion detection systems. IEEE Transactions on Systems, Man, and Cybernetics—Part C: Applications and Reviews. 2008;**38**(5):649-659. DOI: 10.1109/TSMCC.2008.923876

[48] Barbara D, Couto J, Jajodia S, Wu N. ADAM: A testbed for exploring the use of data mining in intrusion detection. In: Proceedings of the ACM SIGMOD. Santa Barbara, CL; May 2001. DOI: 10.1145/604264.604268

[49] Zhang J, Zulkernine M. A hybrid network intrusion detection technique using random forests. In: Proceedings of the 1st International Conference on Availability, Reliability, and Security (ARES'06). Vienna, Austria: IEEE; 20-22 April 2006. DOI: 10.1109/ARES.2006.7

[50] Anderson D, Frivold T, Valdes A. Next-generation intrusion detection expert system (NIDES) – A summary. Technical Report SRI-CSL-95-07, SRI; 1995

[51] Agrawal R, Gehrke J, Gunopulos D, Raghavan P. Automatic subspace clustering of high dimensional data for data mining applications. In: Proceedings of ACM SIGMOD. Seattle, WA: ACM; 1998. pp. 94-105. DOI: 10.1145/276305.276314

[52] Sen J, Sengupta I. Autonomous agent-based distributed fault-tolerant intrusion detection system. In: Proceedings of the 2nd International Conference on Distributed Computing and Internet Technology (ICDCIT'05). Vol. 3186. Bhubaneswar, India: Springer, LNCS; 22-24 December 2005. pp. 125-131. DOI: 10.1007/11604655_16

[53] Sen J, Chowdhury PR, Sengupta I. An intrusion detection framework in wireless ad hoc network. In: Proceedings of the International Conference on Computer and Communication Engineering (ICCCE'06). KL, Malaysia; 10-12 May 2006

[54] Sen J, Sengupta I, Chowdhury PR. An architecture of a distributed intrusion detection system using cooperating agents. In: Proceedings of the International Conference on Computing and Informatics (ICOCI'06). KL, Malaysia: IEEE; 6-8 June 2006. pp. 1-6. DOI: 10.1109/ICOCI.2006.5276474

[55] Sen J. A trust-based detection algorithm of selfish packet dropping nodes in a peer-to-peer wireless mesh network. In: Meghanathan N et al, editors. Recent Trends in Network Security and Applications. CNSA 2010. Communications in Computer and Information Science. Vol. 89. Berlin, Heidelberg: Springer; 2010. pp. 528-537. DOI: 10.1007/978-3-642-14478-3_53

[56] Sen J. A distributed trust and reputation framework for mobile ad hoc networks. In: Meghanathan N et al, editors. Recent Trends in Network Security and Applications. CNSA 2010. Communications in Computer and Information Science. Vol. 89. Berlin, Heidelberg: Springer; 2010. pp. 538-547. DOI: 10.1007/978-3-642-14478-3_54

Chapter 11

Multimodal Biometrics for Person Authentication

Ryszard S. Choras

Abstract

Unimodal biometric systems have limited effectiveness in identifying people, mainly due to their susceptibility to changes in individual biometric features and presentation attacks. The identification of people using multimodal biometric systems attracts the attention of researchers due to their advantages, such as greater recognition efficiency and greater security compared to the unimodal biometric system. To break into the biometric multimodal system, the intruder would have to break into more than one unimodal biometric system. In multimodal biometric systems: The availability of many features means that the multimodal system becomes more reliable. A multimodal biometric system increases security and ensures confidentiality of user data. A multimodal biometric system realizes the merger of decisions taken under individual modalities. If one of the modalities is eliminated, the system can still ensure security, using the remaining. Multimodal systems provide information on the "liveness" of the sample being introduced. In a multimodal system, a fusion of feature vectors and/or decisions developed by each subsystem is carried out, and then the final decision on identification is made on the basis of the vector of features thus obtained. In this chapter, we consider a multimodal biometric system that uses three modalities: dorsal vein, palm print, and periocular.

Keywords: feature transform, multimodal biometric recognition, levels of fusion, dorsal vein, periocular, palm print, PCA

1. Introduction

Biometrics is a technology that uses physical and/or behavioral characteristics of people to identify them. Systems of this type implement two processes (**Figure 1**) [1]:

i. Enrollment

ii. Authentication

The physical features are fingerprints, hand geometry, handprint, facial image, iris, retina, and ear. Behavioral features are signature, lip motion, speech, dynamics of typing, hand movements, and gait.

The characteristics of effective biometrics are:

1. Unique features for each individual

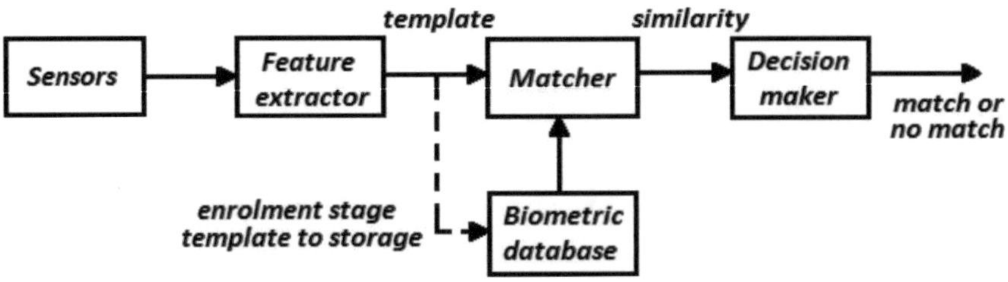

Figure 1.
Biometric recognition system.

2. Invariant traits over time (e.g., due to the effect of aging)
3. Features that are relatively easy to obtain (computational complexity small)
4. Precise algorithms enabling classification
5. Resistance to various types of attacks
6. Low cost
7. Ease of implementation

The security of the biometric system is usually assessed on the basis of some indicators. These are:

- False match rate (FMR). It belongs to the group of matching errors. This indicator is defined as the expected probability that the downloaded sample will be falsely matched to the template in the database, but it will not be the test user pattern. If the indicator is high, it means that there is a risk that an unauthorized person will be recognized as a system user.

- False rejection (FRR) is equivalent to the FMR. The difference between these indicators is that FMR refers to a single match, and the FRR refers to a situation where one or more attempts to match a sample to a template from the database may occur. The FRR error is referred to in the literature as type I error.

- False discrepancy (FNMR). This is the coefficient determining the probability that the sample taken will not be matched to the pattern in the database belonging to the user from whom the sample was taken. In biometric verification (1:1) systems, the indicator means that the sample has not been identified by a specific pattern, while in biometric identification systems (1:N), this indicator determines the probability that a given pattern will not be found in the database.

- The false acceptance factor (FAR) is equivalent to the FNMR indicator. The difference between him and FNMR is the same as between FRR and FMR.

- Equal error rate (EER). It is defined as the intersection of the FAR and FRR characteristics in the graph of the dependence of these errors on the threshold of sensitivity (t). This factor indicates the optimal sensitivity threshold at which the same number of people is incorrectly rejected and incorrectly accepted. The lower the EER error value, the better the biometric system is.

The FMR (FRR) and FNMR (FAR) parameters can also be represented by graphs (**Figure 2**):

- Receiver operating characteristic (ROC) curve showing the dependence of FNMR on FMR. You can use it to show the accuracy of the system.

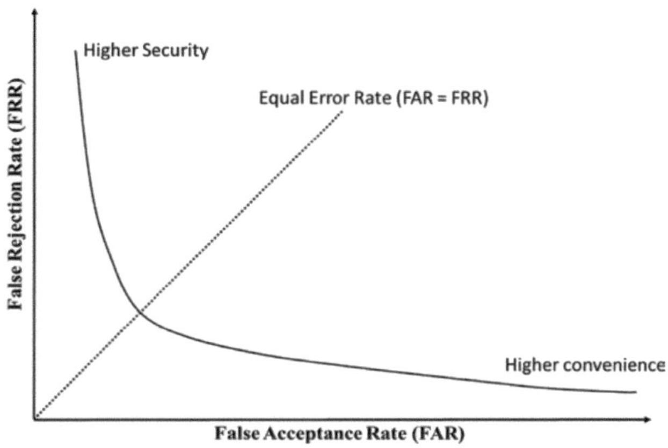

Figure 2.
The graph of FAR, FRR, and EER in receiver operating characteristic (ROC) curve.

Name	Description
Distortion of the input biometric data	Distorted biometric data may prevent the correct alignment process with database templates, as a result of which users are incorrectly rejected or identified
Intra class variations	Biometric data obtained from the person during authentication may differ from the data used to generate the template during registration, thus affecting the matching process. The biometric template should have a small intra-class variance
Interclass similarities	Biometric features should be significantly different for different people and should ensure small similarities between classes in the feature space. There is an upper limit to the users who can be effectively distinguished by any biometric system. The capacity of the identification system cannot be arbitrarily increased for fixed sets of feature vectors and the matching algorithm. The biometric template should have large interclass variations
Non-universality	Obtaining accurate (useful) biometric data from the users is not always possible
Intruder attacks	Attacks of this type involve the manipulation of biometric features to avoid recognition. It is also possible to create artificial biometric patterns in order to accept the identity of another person

Table 1.
Limitations of unimodal biometrics.

Name	Description
Recognition accuracy	The multi-biometric system ensures greater accuracy and reliability thanks to many independent biometric features that are difficult to attack
Continuous monitoring	In case when one biometric modality is obstructed, other modalities of the multi-biometric system ensure correct user identification
Privacy	Multi-biometric systems provide greater resistance to certain types of loopholes and attacks. It is difficult and/or impossible to steal many biometric patterns (templates) stored in the biometric database
Biometric data enrollment	When biometric input data is unavailable or unacceptable by a biometric system, another biometric system modality may be used
Resistance on spoof attacks	Usually the attacker is not able to use many relevant (accurate) spoofed biometrics

Table 2.
Advantages of multi-biometric systems.

- Detection error trade-off (DET) showing error rates on both axes, most often on a logarithmic scale. This curve is plotted for both matching errors and decisions (**Figure 2**).

If we use only one biometric authentication system, the results obtained are not always good enough. Unimodal biometric systems using a single sensor have many limitations, such as lack of uniqueness, universality, and lack of interference level associated with the acquired data, as a result of which they are unable to provide the required level of identification/verification efficiency (**Table 1**). This is due to the fact that the reliability of the biometric modality applied is affected by the precision of a single biometric system (**Table 2**).

2. Multi-biometric systems

2.1 Types of multi-biometric systems

The multi-biometric system can be (**Figure 3**) (a) a multi-sensor system that allows obtaining data from various sensors using one biometric feature, (b) a system with multiple algorithms processing a single biometric feature, (c) a system consolidating multiple occurrences of the same body trait, (d) a system using

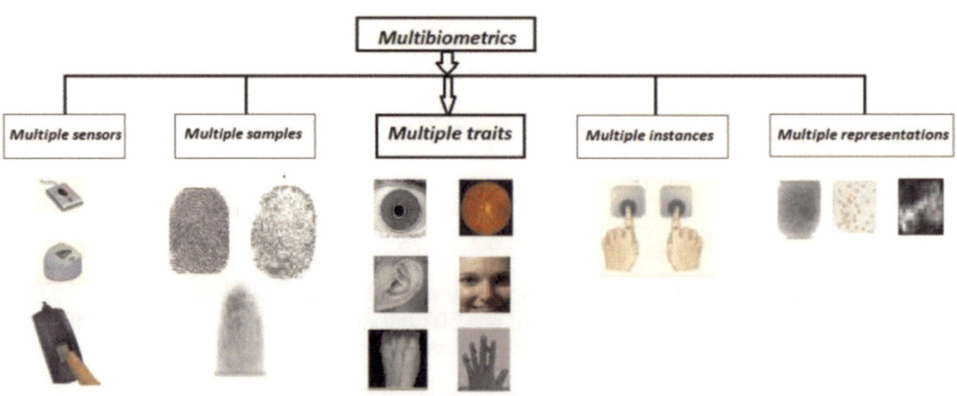

Figure 3.
Types of multi-biometric systems.

multiple templates of the same biometric method obtained with the help of a single sensor, and (e) a multimodal system combining information about the biometric features of the individual to establish his identity [2–4].

2.2 Fusion levels

In multimodal biometric systems, there are a number of strategies (scenarios) for the fusion of biometric information:

- Data fusion from sensors. Data from various sensors form one vector. Fusion of information obtained from many different sensors for a single biometric feature.

- The fusion of feature vectors extracted from various biometric modalities for further processing. A merger of information obtained from several unimodal biometric systems that process different body characteristics of the same person (**Figure 4a**).

- Fusion at the decision level. The merger of decisions developed on the basis of information from different biometric modalities, and the resultant feature vector defines two main classes, i.e., rejection or acceptance (**Figure 4b**).

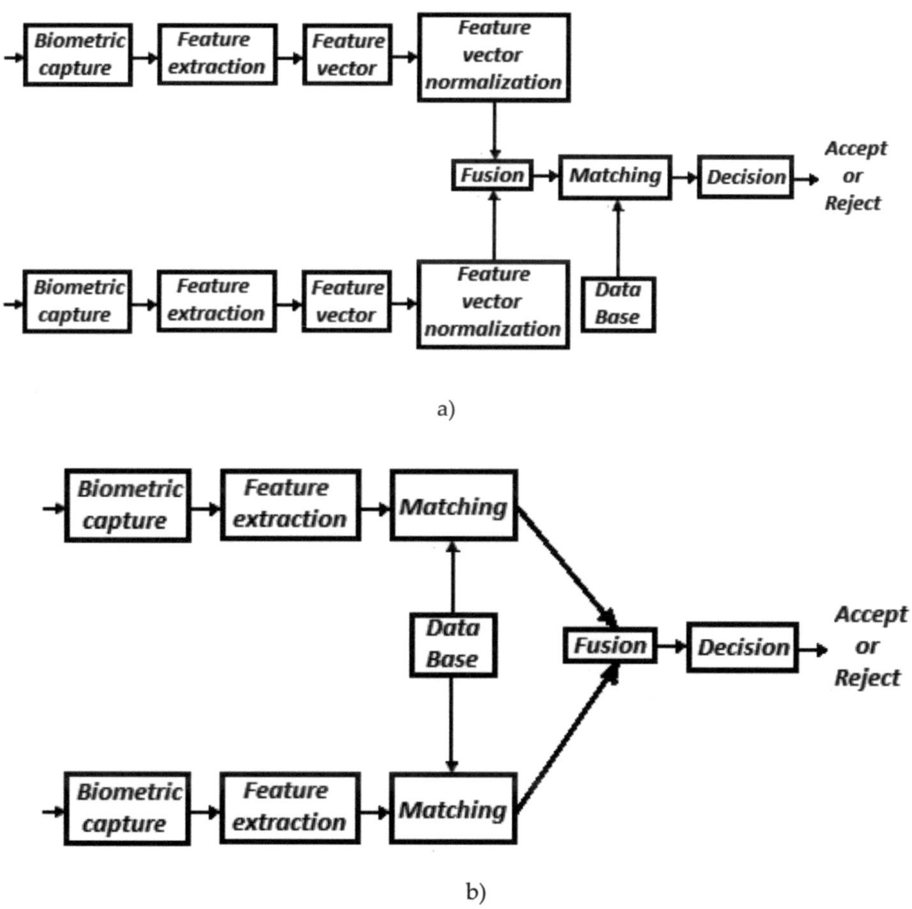

Figure 4.
Levels of fusion. (a) Feature level fusion and (b) score/rank level fusion.

- Rank level fusion. The classifier determines the rank of each registered biometric identity. A high position is a good indicator of a good fit (**Figure 4b**).

2.3 Related work

The fusion of biometrics modalities on different levels of multi-biometrics system is extensively studied in the literature (**Table 3**). For all that the merger at the level of feature vectors is relatively poorly discussed. The merger at this level includes the integration of feature vectors corresponding to many sources of information. Because the feature vectors contain more elements than the input biometric data, it is obvious that the merger at the feature vectors level will provide better

Biometrics traits	**Fusion methods**	**Description of the implementation method**	**References**
Fingerprint and face	**Feature**	In [5], it was proposed to extract face and fingerprint characteristics invariant to the rotation and scaling of Zernike moments (ZM). On the basis of ZM, the fusion of facial features and fingerprints is realized. The RBF network implements the decision-making process. The accuracy rate is 96.55%. Testing result of authentication rate are FAR, 4.95%, and FRR, 1.12%	[5]
	Score	In [6], authors presented score level fusion technique using the SIFT features for the face and the minutiae features for fingerprint. Results are: FAR = 1.98%, FRR = 3.18%, and accuracy = 97.41	[6]
Fingerprint, finger knuckle print, finger vein	**Feature**	The multi-set canonical correlation analysis is used to fuse multiple feature sets. The feature based on MCCA achieves the recognition performance, with EER = 2.3900e-04	[1]
Finger shape		With the help of the unified Gabor filter, fingerprint codes and finger vein codes are generated. The extraction of features is carried out by using a supervised local canonical correlation analysis (SLPCCA), and finally the NN-classifier is used	[7]
Fingerprint and iris	Score	In [8], authors propose a frequency approach to generate a unified homogeneous template for fingerprint and iris features. Scores generated from these templates are fused using the sum rule	[8]
Palm print and hand shape	**Feature**	Information from the face image and gait image are combined at the function level. Using the principal component analysis (PCA) method, facial features were obtained The result of multiple discrimination analysis (MDA) is gait energy image (GEI) Recognition rate results are 91.3%	[9]
Palm print and iris	**Feature**	In system described in [10], texture parameters are extracted based on Gabor filters.	[10]
		Fusion of the palm print features and iris features is based on the wavelets. Decision is obtained using kNN classifier. Recognition accuracy is 99.2% and FRR = 1.6%	
		In [11], fusion method for the information of phases about the iris and palm utilizes a Baud limited image product (BLIP)	[11]
Finger knuckle and palm print	**Feature**	In this paper, feature extraction method for palm print is monogenic binary coding; for inner knuckle print recognition, two algorithms named ridgelet transform	[12]

Biometrics traits	Fusion methods	Description of the implementation method	References
		and SIFT are proposed. The extracted feature vectors are classified using SVM	
Palm print and face	**Feature**	The PCA is used to extract features of palm and face images. Fusion technique concatenated the feature vectors of the face and palm modalities into one fused vector, and feature selection is performed.	[13]
Face and gait	**Feature**	Method is based on learning face and gait features in image transform spaces. Two methods are considered—PCA and LDA	[14]
Face and iris	Score	Multi-biometrics system using dual iris, visible and thermal face traits is considered. 1D Log-Gabor and Complex Gabor Jet Descriptor (CGJD) were used to extract feature vectors. Authors proposed a score level fusion algorithm	[15]
		The ordinal measures and local binary pattern (LBP) methods are proposed to extract features from iris and face regions, respectively	[16]
	Feature	Paper [17] presents the extractions of iris features based on 2D Gabor and facial features using the PCA method	[17]
Face and hand geometry	**Feature**	The 2D DCT is used to extract discriminant face features which are concatenated with hand geometric features. The resultant feature vector is classified using SVM	[18]
Face and ear	Scores	To match score level, fusion is proposed in [19]. Authors use Dempster-Shafer decision theory for each modality. Recognition rate is 95.53% with 4.47% EER	[19]
Ocular—iris and conjunctival vascular	Score	In [4], authors presented fusion of both iris and conjunctival vascular information. A weighted fusion method is proposed for each modality. The fusion resulted in an EER of 2.83%	[4]
Face, ear, and signature	Rank	In [20], the PCA and Fisher's linear discriminant (FLD) methods in the face, ear, and signature, multimodal biometrics system is proposed. Local features are extracted from face, ear, and signature data. Features are matched using Euclidean distance. This system is using rank level fusion	[20]

Table 3.
Summary of works on multimodal biometric systems.

authentication results. However, mergers at this level are difficult to implement in practice because (i) sets of features of many modalities may be incompatible, (ii) the combination of two feature vectors may result in a vector of features with very large dimensionality, and (iii) a complex comparing system is required.

3. The proposed multi-biometric system

The multi-biometric system (dorsal vein + periocular + palm print) is presented in **Figure 5**.

In our proposed method, the first is preprocessing block including noise elimination, ROI detection and normalization, and contrast normalization. For all three

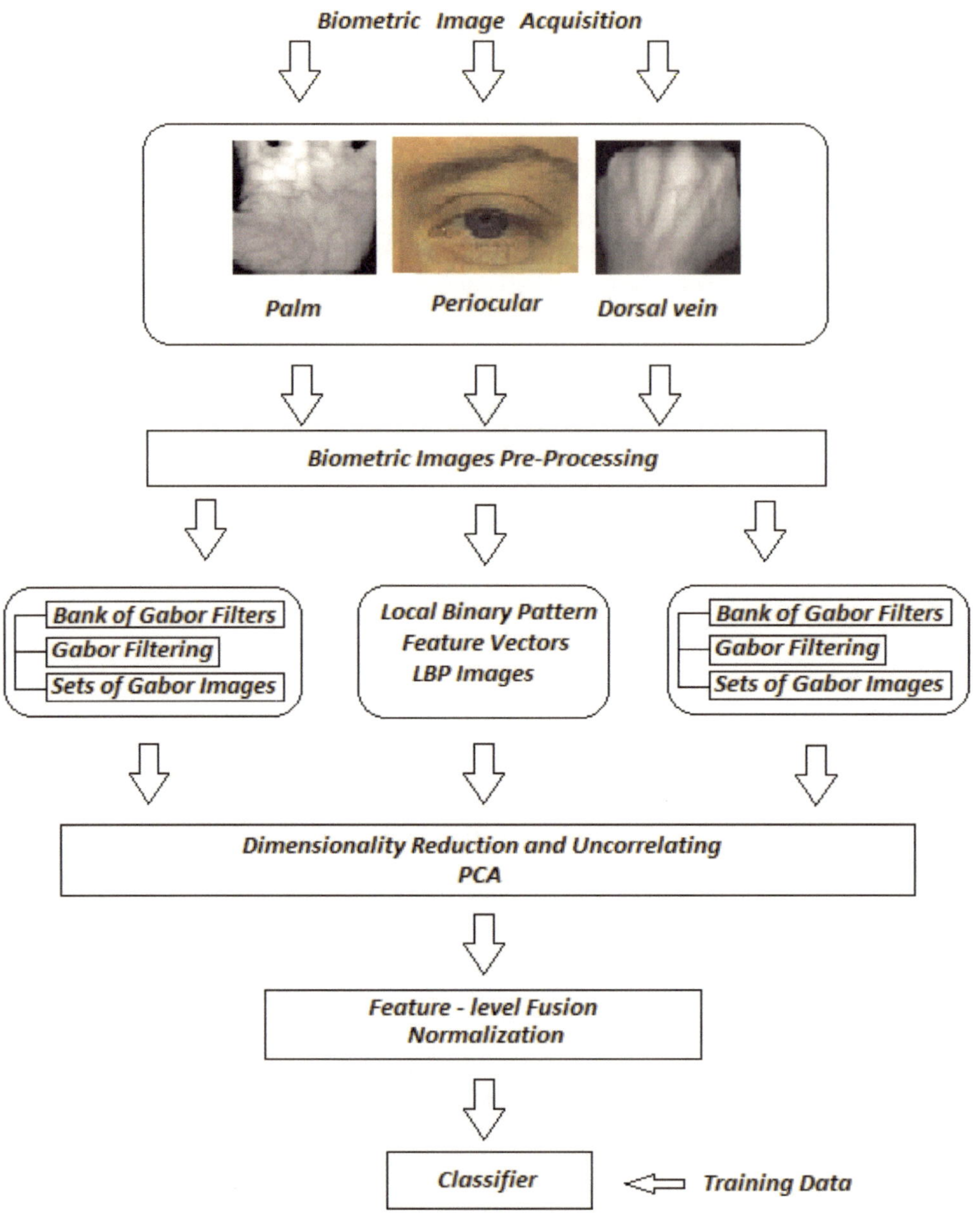

Figure 5.
Considered multi-biometric system architecture.

modalities, noise elimination for an image $f(x,y)$ is performed using median filtering (2D MF) operation formulated as [21]:

$$\hat{f}(x,y) = median_{A_1} f(x,y) = median[f(x+r, y+s)] \tag{1}$$

where $A_1$is the *MF* window.

Next step in preprocessing phase is ROI detection and normalization (**Figure 6**). This operation is quite different for dorsal vein images, palm print images, and periocular images. For dorsal vein images, we use distance transform to detect the dorsal image center and build square ROI based on this center coordinates [22, 23]. The ROI design for palm print images is based on hand-specific points (finger valleys) and two angles [24]. The periocular region is detected based on the center of the iris. Using the conventional algorithm for detecting the iris, we determine the center of the iris and its diameter. The periocular area is a rectangle centered in the iris center [25, 26].

After the ROI detection, we perform image size normalization and apply the contrast normalization by using CLAHE algorithm. The image is divided into non-overlapping areas of equal size, and the histograms of each region are calculated. Next, the cutoff threshold for histograms is obtained, and each histogram is processed in such a way that its height does not exceed the cutoff threshold [21].

The sample input images after normalization operations and operations using the CLAHE algorithm are shown in **Figure** 7. Next processing blocks include feature extraction, feature selection, fusion, and classification.

3.1 Gabor feature extraction

In biologically inspired vision models, receptor fields exist that are the primary aspect of early visual processing in mammalian vision systems. Gabor functions are widely used in image feature analysis because they are similar to receptive field profiles in mammalian cortical simple cells. These fields are modeled using Gabor filters [27].

Imitation of mammalian vision systems (or some of them) in object recognition systems leads to their increased efficiency and plausibility. Object recognition systems that are inspired by the biological approach use filter banks, in particular Gabor filters (**Figure 8**) [28–32].

The 2D Gabor filter family can be represented as expressed in Eq. (2):

$$Gab_{\omega,\theta}(x,y) = \frac{1}{2\pi\sigma_x\sigma_y} G_\theta(x,y) S_{\omega,\theta}(x,y) \tag{2}$$

where $G_\theta(x,y) = e^{-\left(\frac{(xcos\theta+ysin\theta)^2}{2\sigma_x^2}+\frac{(-sin\theta+ycos\theta)^2}{2\sigma_y^2}\right)}$ and $S_{\omega,\theta}(x,y) = e^{i(\omega xcos\theta+\omega ysin\theta)} - e^{-\frac{\omega^2\sigma^2}{2}}$.

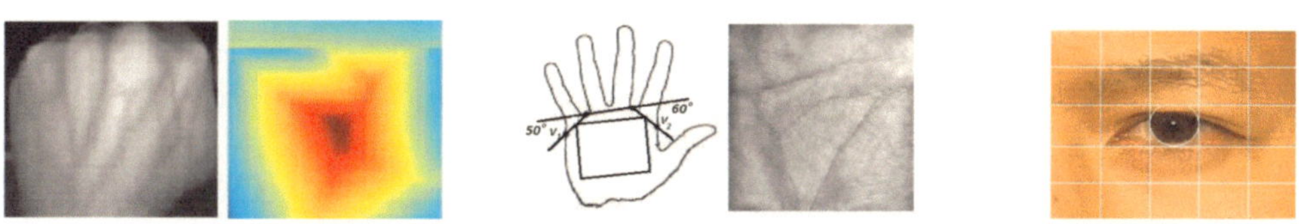

a) b) c)

Figure 6.
ROI area for dorsal vein images (a), palm print images (b), and periocular images (c).

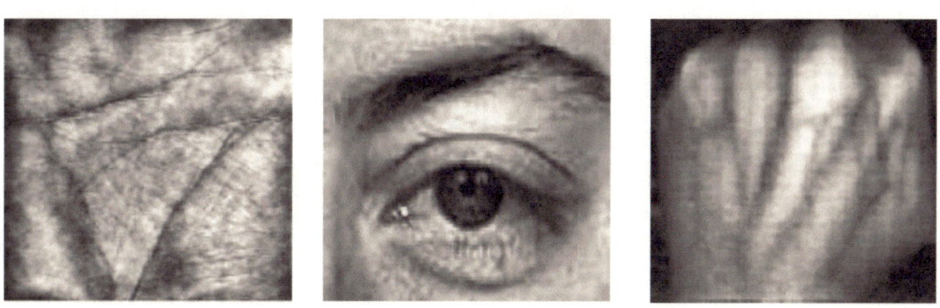

Figure 7.
Images after normalization (size 150 × 150 pixels) and after applying the CLAHE algorithm.

The $Gab_{\omega,\theta}(x,y)$ can be decomposed into a real $\Re\{Gab_{\omega,\theta}(x,y)\} = \frac{1}{2\pi\sigma^2} G_\theta(x,y)\Re\{S_{\omega,\theta}(x,y)\}$ and an imaginary.
$\Im\{Gab_{\omega,\theta}(x,y)\} = \frac{1}{2\pi\sigma^2} G_\theta(x,y)\Im\{S_{\omega,\theta}(x,y)\}$ parts (for $\sigma_x = \sigma_y = \sigma$).

Gabor response images are obtained by convolution operation of multiscale and multi-orientation Gabor filters $Gab_{\omega,\theta}(x,y)$ with the image *f (x, y)*.

$$G_{\omega,\theta}(x,y) = f(x,y) * Gab_{\omega,\theta}(x,y) = Mag_{\omega,\theta}(x,y)e^{i\, Ph_{\omega,\theta}(x,y)} \tag{3}$$

$$Mag_{\omega,\theta}(x,y) = \sqrt{\Re\{Gab_{\omega,\theta}(x,y)\}^2 + \Im\{Gab_{\omega,\theta}(x,y)\}^2},$$

$$Ph_{\omega,\theta}(x,y) = \arctan \frac{\Im\{Gab_{\omega,\theta}(x,y)\}}{\Re\{Gab_{\omega,\theta}(x,y)\}},$$

where and $*$ is the convolution operator.

The Gabor filter responses for palm print image and dorsal vein image are shown in **Figures 9** and **10**, respectively.

3.2 Periocular feature extraction by LBP

The periocular area contains the iris, eyes, eyelids, eyelashes, and partially eyebrows. The *LBP* method can be used to describe the texture of the periocular

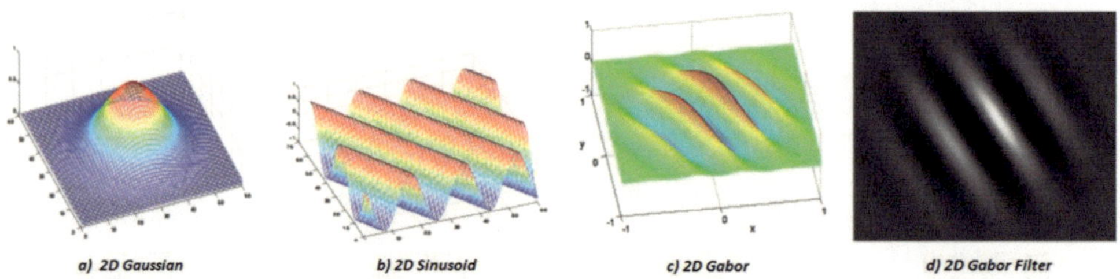

Figure 8.
2D functions and 2D Gabor filter.

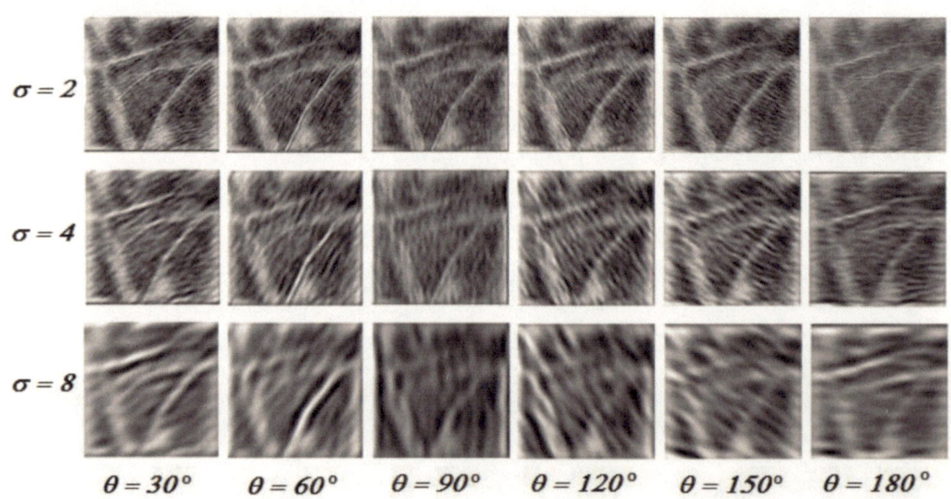

Figure 9.
Imaginary part of the Gabor filter responses of a palm print image.

area, and the feature vectors contain LBP features. The operator of local binary patterns (*LBP*) was proposed by Ojala [33] as a texture descriptor.

LBP divides the image into non-overlapping blocks of the same size. Local image features are calculated for each block separately. For a set of pixels belonging to a given block, the *LBP* values are calculated and then a histogram is created. The feature vectors (histograms) of each block are combined to form a global vector of features of the entire image.

LBP analyzes the local neighborhood consisting of g_p points located on a circle with radius R and surrounding the center point of g_c and checks whether the points of g_p are greater or lesser than the g_c point value.

The *LBP* value of the g_c point is specified as follows:

$$LBP_{P,R} = \sum_{p=0}^{P-1} S\left(g_p - g_c\right) 2^p \tag{4}$$

where g_p and g_c are the luminance values of the neighborhood and center point, respectively.

The idea of this operator is presented in **Figure 11**.

For an image size $M \times N$, the image descriptor is a histogram created from the *LBP* values:

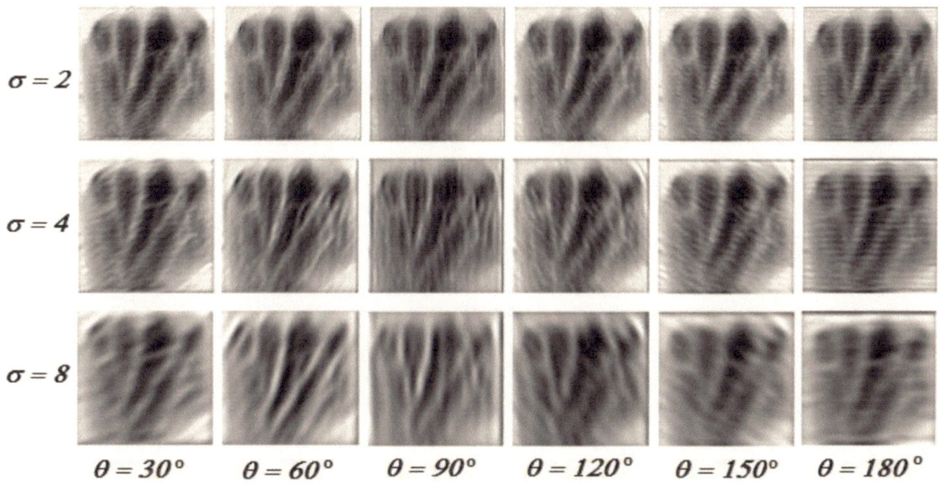

Figure 10.
Imaginary part of the Gabor filter responses of a dorsal vein image.

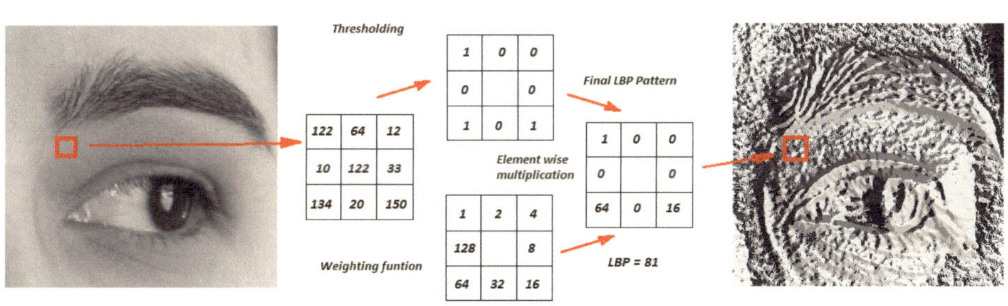

Figure 11.
The basic idea of LBP approach.

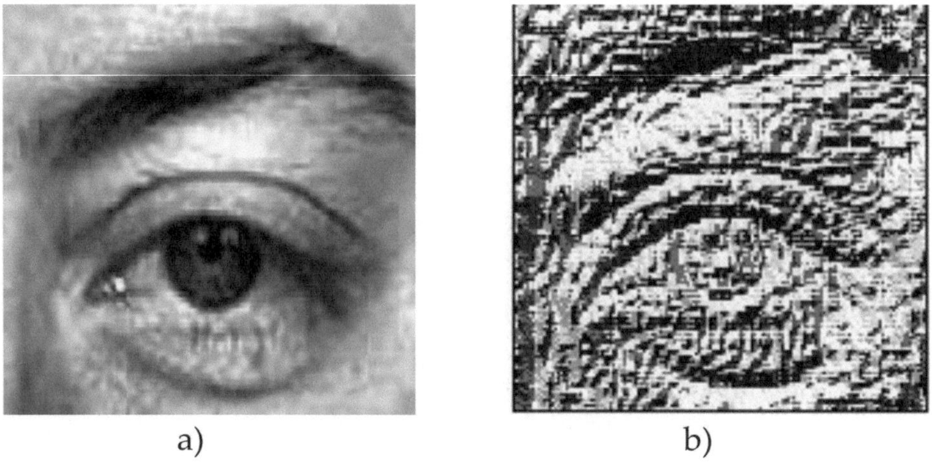

Figure 12.
The original image (a) and image as a result of the LBP operator (b).

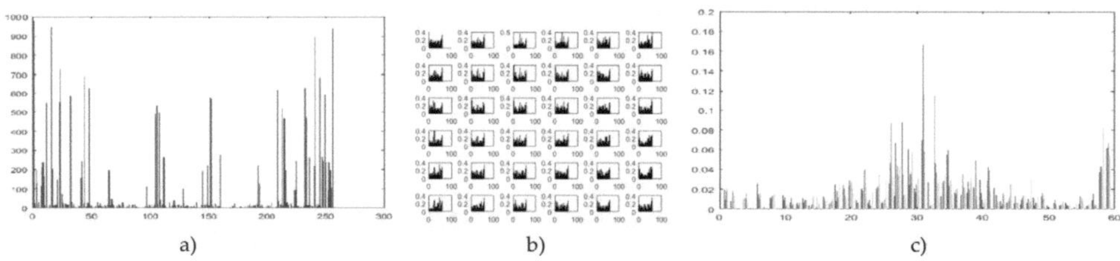

Figure 13.
The $LBP_{P,R}$ histogram (a), histograms of the n *blocks (b), and the $LBP^{u2}_{P,R}$ histogram (c).*

$$H(k) = \sum_{i=1}^{M} \sum_{j=1}^{N} f(LBP_{P,R}(i,j), k); k \in [0, K] \quad (5)$$

$$f(x, y) = \begin{cases} 1, & x = y \\ 0, & \text{otherwise} \end{cases}$$

where k is one LBP pattern and K is the maximal LBP pattern value (number bin of the histogram).

Using the LBP operator, we obtain 2^P different output values corresponding to 2^P different binary patterns created by P of neighboring pixels. Certain binary patterns contain more information than others, so we can only consider this subset of LBP values. Patterns of this subset are called uniform patterns. So we have a standard $LBP_{P,R}$ operator and an $LBP^{u2}_{P,R}$ operator.

Typically image is divided into n blocks and histograms of each block are concatenate into feature vector [34].

In the case $LBP_{P,R}$ operator, the histogram contains 256 bins. In the case *of* $LBP^{u2}_{P,R}$ operator, the histogram contains 59 bins (**Figures 12** and **13**).

4. Feature reduction and data fusion

The multi-biometric system has been tested using certain parts of the following databases: PolyU palmprint [24], IIITD periocular database [25], and Bosphorus hand vein database [35]. We choose 20 subjects with 10 images per subject at random. From 10 images, 5 images are used for training and 5 for the testing.

The combination of feature vectors at this level is difficult to achieve in practice due to the combination of certain fundamentally different feature vectors that can result in a resulting vector of features with very large dimensionality. In a merger at the level of feature vectors, each individual modality process generates a feature vector. The fusion process combines these feature vectors into one vector.

For dorsal vein images and for palm print images, we perform the same image processing operations that the feature vectors have the same sizes. As a result of convolution operation of multiscale and multi-orientation Gabor filters with the input image, we get the Gabor response images. The feature vector has a very large size of *(M x N x k x l)* where *M x N* is the image size, *k* is the number of scales, and *l* is the number of orientations. In our case, for both dorsal vein images and palm print images, we get a feature vector containing $(150 \times 150 \times 3 \times 6) = 405{,}000$ items. The images subjected to the Gabor filtration are rescaled with a scale factor of 0.1, which allows obtaining a vector of features with a size of 1×4050 elements.

For periocular images, the feature vector has a size of $36 \times 59 = 2124$.

Next we reduce dimensionality of these vectors used in PCA method (**Figure 14** and **Table 4**) [5]. Separated features are normalized using zero mean and unit variance as

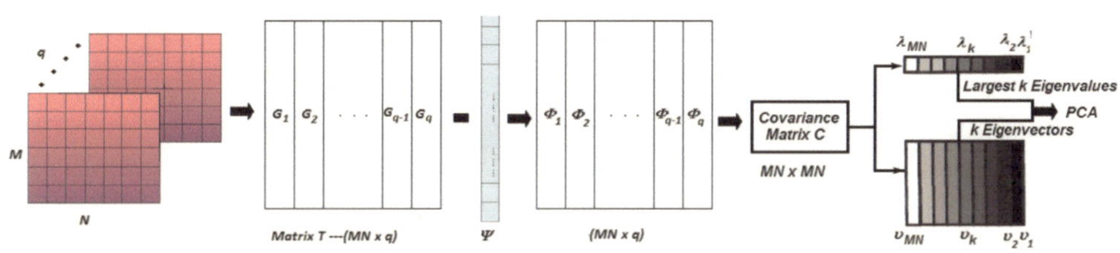

Figure 14.
Steps to image processing using PCA.

PCA algorithm
Organizing the training set of images $T = [G_1, G_2, \cdots, G_q]$ where q is the number of images in the training set
Calculating the average of the set T $\Psi = \frac{1}{q}\sum_1^q G_q$
Calculating $\Phi_i = G_i - \Psi$
Calculating the covariance matrix C $C = \frac{1}{q}\sum_1^q \Phi_i \Phi_i^t = AA^t$
The eigenvectors and corresponding eigenvalues are computed $C v_i = \lambda_i v_i i = 1, \cdots, q$
The eigenvectors and their corresponding eigenvalues are paired and ordered from high to low. Approximated image is calculated as $\overline{G} = vw + \Psi$ for $v = (v_1, v_2, \cdots, v_k)$

Table 4.
PCA algorithm.

Modality	Number of the eigenvectors			
	k = 40	k = 60	k = 80	k = 100
Dorsal vein	88	89.3	91.4	92.6
Palm print	88.7	89.3	90.6	92.8
Periocular	86	86.8	89	89.2
Dorsal vein + palm print	90.3	91.1	92.3	93.1
Dorsal vein + periocular	91.1	92	92.4	92.8
Palm print + periocular	90.7	91.4	91.8	92.1
Dorsal vein + periocular + palm print	**93.2**	**94**	**94.5**	**95.3**

Table 5.
Recognition rates [%] for different modality.

$$\bar{f}_i = \frac{f_i - \mu_i}{\sigma_i} \quad (6)$$

where μ_i and σ_i are the mean value and standard deviation of the i-th feature, $\bar{f}_i$ is the normalized i-th feature vector.

Table 5 shows the recognition performance depending on the number of selected eigenvectors.

5. Conclusion

In this chapter, Gabor's functions and LBP features are proposed for recognition in a multi-biometric system that uses three modalities: dorsal vein, periocular, and palm print. Using PCA method dimensionality feature vectors from these modality are reduced. Feature vectors are normalized and fused using concatenation operation. Based on the results, we suggest that multi-biometric system using the fusion of dorsal vein, periocular, and palm print images can offer recognition rate which the unimodal biometric system cannot.

Author details

Ryszard S. Choras
Institute of Telecommunications and Computer Sciences, UTP University of Science and Technology, Bydgoszcz, Poland

*Address all correspondence to: choras@utp.edu.pl

References

[1] Ross AA, Nandakumar K, Jain AK. Handbook of Multibiometrics. Boston, MA, USA: Kluwer; 2006

[2] Travieso CM, del Pozo-Baños M, Alonso JB. Fused intra-bimodal face verification approach based on scale-invariant feature transform and a vocabulary tree. Pattern Recognition Letters;**36**:254-260

[3] Travieso CM, Zhang J, Miller P, Alonso JB, Ferrer MA. Bimodal biometric verification based on face and lips. Neurocomputing;**74**(14–15): 2407-2410

[4] Ross A. An introduction to multibiometrics. In: Proceedings of the 15th European Signal Processing Conference; 2007. pp. 20-24

[5] Long TB, Thai LH, Hanh T. Multimodal biometric person authentication using fingerprint, face features. In: Anthony P, Ishizuka M, Lukose D, editors. Trends in Artificial Intelligence (LNCS 7458). Springer; 2012. pp. 613-624

[6] Choi H, Choi K, Kim J. Mosaicing touchless and mirror-reflected fingerprint images. IEEE Transactions on Information Forensics and Security. 2010;**5**(1):52-61

[7] Ghouti L, Bahjat AA. Iris fusion for multibiometric systems. In: Proceedings of the IEEE International Symposium on Signal Processing and Information Technology; 2009. pp. 248-253

[8] Ross A, Jain A. Information fusion in biometrics. Pattern Recognition Letters. 2003;**24**(13):2115-2125

[9] Chen Y, Parziale G, Diaz-Santana E, Jain AK. 3D touchless fingerprints: Compatibility with legacy rolled images. In: Proceedings of the Biometrics Symposium: Special Session on Research at Biometric Consortium Conference; 2006. pp. 1-6

[10] Froba B, Rothe C, Kublbeck C. Evaluation of sensor calibration in a biometric person recognition framework based on sensor fusion. In: Proceedings of 4th IEEE International Conference on Automatic Face & Gesture Recognition; 2000. pp. 512-517

[11] Meraoumia A, Chitroub S, Bouridane A. Multimodal biometric person recognition system based on fingerprint & finger-knuckleprint using correlation filter classifier. In: Proceedings of the IEEE International Conference on Communications; 2012. pp. 820-824

[12] Bhaskar B. Veluchamy S. Hand based multibiometric authentication using local feature extraction. In: Proceedings of the International Conference on Recent Trends in Information Technology; 2014. pp. 1-5

[13] Bokade GU, Sapkal AM. Feature level fusion of palm and face for secure recognition. International Journal of Electrical and Computer Engineering. 2012;**4**(2):157

[14] Hossain E, Chetty G. Multimodal face-gait fusion for biometric person authentication. In: Proceedings of the IFIP 9th International Conference on Embedded Ubiquitous Computing; 2011. pp. 332-337

[15] Ding Y, Zhuang D, Wang K. A study of hand vein recognition method. In: Proceedings of the IEEE International Conference on Mechatronics & Automation; 2005. pp. 2106-2110

[16] Miao D, Sun Z, Huang Y. Fusion of multibiometrics based on a new robust linear programming. In: Proceedings of the 22nd International Conference on Pattern Recognition; 2014. pp. 291-296

[17] Shah S, Ross A, Shah J, Crihalmeanu S. Fingerprint mosaicking using thin plate splines. In: Proceedings of the Biometric Consortium Conference; 2005. pp. 1-2

[18] El-Alfy E-SM, BinMakhashen GM. Improved personal identification using face and hand geometry fusion and support vector machines. In: Benlamri R, editor. Networked Digital Technologies. Vol. 294. Springer; 2012. pp. 253-261

[19] Jing X-Y, Yao Y-F, Zhang D, Yang J-Y, Li M. Face and palmprint pixel level fusion and kernel DCV-RBF classifier for small sample biometric recognition. Pattern Recognition. 2007;**40**(11): 3209-3224

[20] Rattani A, Freni B, Marcialis GL, Roli F. Template update methods in adaptive biometric systems: A critical review. In: Tistarelli M, Nixon MS, editors. Advances in Biometrics (LNCS 5558). Springer; 2009. pp. 847-856

[21] Choras, R.S., A survey on methods of image processing and recognition for personal identification. In: Machine Learning and Biometrics, Rijeka, Croatia: InTech; 2018

[22] Tanaka T, Kubo N. Biometric authentication by hand vein patterns. In: Proceedings of the SICE Annual Conference; 2004. pp. 249–253

[23] Ferrer MA, Morales A, Travieso CM, Alonso JB. Low cost multimodal biometric identification system based on hand geometry, palm and finger print texture. In: 41st Annual IEEE International Carnahan Conference on Security Technology; 2007. pp. 52-58

[24] Zhang D, Kong A, You J, Wong M. Online palmprint identification. IEEE Transactions on Pattern Analysis and Machine Intelligence;**25**:1041-1050

[25] Sharma A, Verma S, Vatsa M, Singh R. On cross spectral periocular recognition. In: Proceedings of International Conference on Image Processing; 2014

[26] Woodard D, Pundlik S, Lyle J, Miller P. Periocular region appearance cues for biometric identification. In: Computer Vision and Pattern Recognition Workshops; 2010. pp. 162-169

[27] Gabor D. Theory of communication. The Journal of The Institute of Electrical Engineers of Japan. 1946;**93**:429-459

[28] Wang N, Li Q, El-Latif AAA, Yan X, Niu X. A novel hybrid multibiometrics based on the fusion of dual iris, visible and thermal face images. In: Proceedings International Symposium on Biometrics and Security Technology; 2013. pp. 217-223

[29] Choras RS. Image feature extraction techniques and their applications for CBIR and biometrics systems. International Journal of Biology and Biomedical Engineering. 2007;**1**(1):6-16

[30] Choras RS. Iris-based person identification using Gabor wavelets and moments. In: Proceedings of the International Conference on Biometrics and Kansei Engineering; 2009. pp. 55-59

[31] Choras RS. Personal identification using forearm vein patterns. In: International Conference and Workshop on Bioinspired Intelligence; 2017. pp. 1-5

[32] Choras RS. Biometric personal authentication using images of forearm vein patterns. In: International Conference on Signals and Systems; 2017. pp. 40-43

[33] Ojala T, Pietikäinen M, Mäenpää T. Multiresolution gray-scale and rotation invariant texture classification with local binary patterns. IEEE Transactions

on Pattern Analysis and Machine Intelligence. 2002;**24**(7):971-987

[34] Wang Y, Li K, Cui J. Hand-dorsa vein recognition based on partition local binary pattern. In: IEEE International Conference on Signal Processing; 2010. pp. 1671-1674

[35] Yüksel A, Akarun L, Sankur, B. Biometric identification through hand vein patterns. In: ICPR'2010: International Conference on Pattern Recognition; Istanbul; 2010

Chapter 12

Hardware Implementation of Audio Watermarking Based on DWT Transform

Amit M. Joshi

Abstract

Presently, the duplicate copy of an audio can be generated with great ease using some smart devices, and transmitted over the internet which raises concern over copyright and privacy. Digital audio watermarking is a procedure to insert some data bits known as watermark into audio signal. Then the audio with watermark is to be transmitted to end user or made public. The proposed algorithm is used to insert a binary watermark image into a detailed coefficient of the Daubechies 9/7-based DWT transform. A watermark is dispersed consistently in low frequencies, which builds the robustness and inaudibility of the watermark data. Further, the watermark is embedded into an audio signal to have robust system against audio attacks and inaudible performance. The algorithm is verified using MATLAB and subsequently implemented on FPGA hardware to verify the real-time performance. Hardware implementation helps to embed the watermark at the same instance when audio is being captured. The results show promising application for real-time audio applications.

Keywords: audio, digital watermark, FPGA, real-time, robustness

1. Introduction

In a present digital era, a digital file like audio can be copied easily to a computer and other smart devices, and distributed on open network. However, this has prompted issues such as maintaining copyright, ownership, particular person authentication, privacy, and sensitive information loss [1]. The possible solution is to insert some ownership data bits into the audio which would be extracted for the purpose of the authentication. Digital audio watermarking is a technique where a watermark is embedded in the original audio media file. Subsequently, the secured watermarked may be transmitted over internet to any other person. Inaudibility and robustness are two primary characteristics of a digital audio watermark. Robustness is defined as the ability of watermark to resist channel attacks like echo addition, filtering and Gaussian noise, etc. [2]. Inaudibility means the insertion of the watermark should not have any impact on final watermarked audio. Ownership protection helps to identify the content for the originator in order to protect his copyrights. Illegal use of audio without consent, leaking sensitive information, etc. can be prohibited by embedding owner signature into original audio in real-time [3].

The main objective is to design an algorithm which is robust, blind and inaudible and useful for audio applications.

1.1 Digital watermarking overview

The following section gives a brief overview of digital watermarking. Some basic terms, watermark classifications, watermark properties, and applications covered under this chapter. The following list contains the meaning of some standard terms used in this chapter.

- Host audio is the source audio signal.
- Watermark is defined as a signal consisting of data embedded into a host/ carrier audio signal.
- Watermark Embedding is the process of inserting the ownership data into host audio.
- Blind Watermarking is a technique in which there is no need of source audio for watermark extraction.
- Watermark extraction is a procedure to retrieve back to our embed watermark.
- The payload is the size of the message encoded in object [4].

1.2 Problem statement

There are so many audio watermarking algorithms which are implemented in previous year. Most of the algorithms are implemented on the MATLAB only and then it checks its robustness and inaudibility. In MATLAB, the transform function is generally used and according to that an audio watermarking algorithm is applied to frequency domain. In MATLAB, the transform function is generally used and according to that an audio watermarking algorithm is applied in the frequency domain. The power consumption is also unknown and also do not have any knowledge about execution time of the algorithm. These are the some fundamental requirement to design any algorithm on hardware so MATLAB does not provide any kind of hardware support. The hardware implementation of algorithms are achieved on DSP processor and also on GPU processor level. DSP processor and GPU processor may give hardware implementation but its hardware complexity is very high and they are not compatible with the real-time applications [5]. So, VLSI architecture is the best suitable platform for reducing hardware complexity and designing on real-time applications.

1.3 Objectives

The Proposed design of audio watermarking algorithm is implemented on MATLAB. Subsequently VLSI architecture of the audio watermarking algorithm is developed. Then a Forward DWT transform algorithm is developed in Xilinx ISE which is followed by inverse DWT algorithm. Then design VLSI architecture of blind audio watermarking algorithm is developed. Here the main objectives of this proposed work is design VLSI architecture of the blind audio watermarking algorithm and also check its area and timing calculation. The proposed work is also designed to have compatibility with real-time application.

1.4 Previous work and my contribution

Digital audio watermarking is used for correct owner identification, prevention of fragile and copying and also providing a particular person authentication of their digital property. There are many digital audio watermarking algorithms are designed and simulated on MATLAB platform. So many types of audio watermarking methods present in a previous year [6–8]. Also, there is a DWT SVD-based audio watermarking algorithm is implemented in previous work [9]. This work based on semi-blind audio watermarking-based algorithm and a digital watermark is applied on DWT-SVD transform with robustness and imperceptible. The proposed algorithm is a blind digital audio watermarking scheme using DWT algorithm. There are several hardware implementation of the DWT algorithm [10–12]. In the proposed algorithm, the reduced the complexity of the DWT is designed along with its inverse DWT algorithm. The real-time application requires high speed of the algorithm. Our algorithm gives less delay with complete synchronization which does not require any control segment as suggested by many scholars which increase delay. Here, the hardware implementation uses only adders, subtractors and shifters so multiplier-less designed would help to have the hardware efficient and very fast algorithm.

1.5 Hardware solution

The scheme of watermarking is implemented either using software or hardware. In a software implementation, a watermark algorithm is executed on a processor. The software implementation is flexible, but the software implementation is used to embed watermark on offline process where algorithm runs on PC for audio captured through the device. However, the hardware implementation helps to insert the watermark online when the audio is being recorded itself. Then again, in a hardware implementation, a watermark calculation is entirely performed in specially crafted hardware. A hardware implementation consumes less area and less power contrasted with a software implementation [5]. The hardware implementation may have the advantage of parallel processing and poses lesser delay compared to software. This chapter is targeting a real time application, so hardware solution is best recommended. Initially, the proposed audio watermarking algorithm is implemented on the MATLAB; however MATLAB provides only the simulation platform to validate the performance [13]. The real-time implementation of the proposed audio watermarking is achieved in Xilinx ISE software and simulated result of the audio watermarking is discussed. Here DWT transform is implemented by using adder/subtractor and shifter only. Then steps of both embedded and extraction process of the digital audio watermarking is implemented. Subsequently, the proposed watermarking is also synthesized using Xilinx ISE14.7.

2. Digital watermarking

Watermarking is a method through which the protected data conveyed without much observable change in the watermarked content. The watermarking process comprises of two main steps: (i) embedding method and (ii) extraction method. The secret key could be used for additional level of security. There are fundamentally three sorts of watermarking methods and are described here: (1) non-blind watermarking, (2) semi-blind and (3) blind watermarking. The process of watermarking that uses the original sound signal during the extraction procedure termed as "non-blind watermarking". The watermarking system uses a portion of the segment or some a part of the input audio signal then it is term as a "semi-blind watermarking". The watermarking system helps to retrieve the watermark without use of original audio signal or a part of an audio signal for extraction process termed as "blind watermarking" [14]. The paper covers proposed novel blind audio watermarking scheme and its hardware implementation is performed in Xilinx ISE. The steps of algorithms are covered in Section 3. The watermark consists of a data sequence of binary bits which is inserted into the host signal. The audio watermarking scheme should have following basic characteristics: inaudibility, payload and robustness. **Figure 1** gives the visual representation of the requirements of data watermarking concept in digital audio, these three requirements forms the corners of the magic triangle.

1. **Inaudibility:** The inserted data has to be "inaudible" in the watermarked digital music. Evaluation of the same is quantified using signal-to-noise ratio (SNR).

2. **Security:** The algorithm should be secure where authorized person should able to only retrieve the watermark. The attempt of extracting watermark is to be unsuccessful for an unauthorized user in any case.

3. **Robustness:** The watermark should not be eliminated or removed by applying common processing techniques such as cropping, nonlinear and/or linear filter, lossy compression, etc.

4. **Paylod (capacity):** It is defined as total information to be embedded in the host without having of any distortion. It is usually defined as the bit rate for the audio signal which is the actual number of bits inserted in the original audio and is provided by bits per second (bps).

5. **Real-time processing:** The process of inserting into the original signal without much delay. It should be able to insert the watermark at same instance when the audio is being recorded.

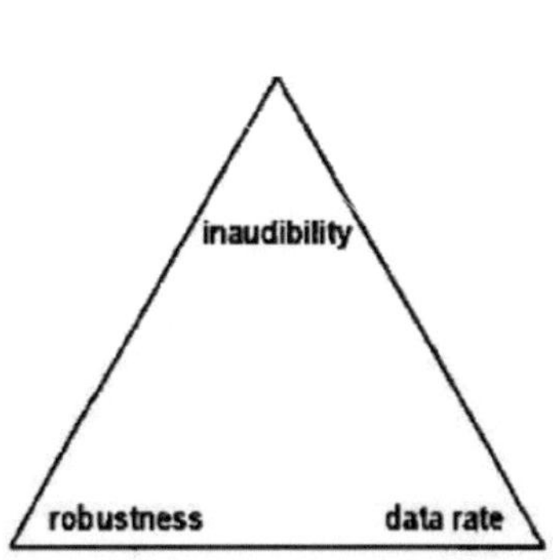

Figure 1.
Performance parameter magic triangle.

3. Proposed audio watermarking

The proposed audio watermarking scheme is blind and robust and is based on DWT transformation. In the proposed scheme, only eight-audio samples of a single frame from two channels, is considered for watermarking process. The details of the embedded and extracted process are shown in the following section.

3.1 Embedding process of DWT-based audio watermarking

Input: original audio, watermark; output: watermarked audio.

The steps of embedding process are as follows:

Step 1: The original audio signal of 16 samples is considered from two channels for further watermark embedding process.

Step 2: DWT transform is applied to obtain an approximate and detailed coefficient of the both channels. Here the approximate and detailed coefficients are low and high pass filter component of the original input signal.

Step 3: Then, binary bits are embedded in the detailed component of an input audio signal. If watermark bit is one then according to

$$P1' = P1 + (I^* P1) \text{ in the first channel} \tag{1}$$

where P_1' = detailed component after watermarking, P_1 = detailed component before watermarking, I = intensity factor and if watermark bit is 0, then 2nd channel detailed component is changed with the first channel detailed component.

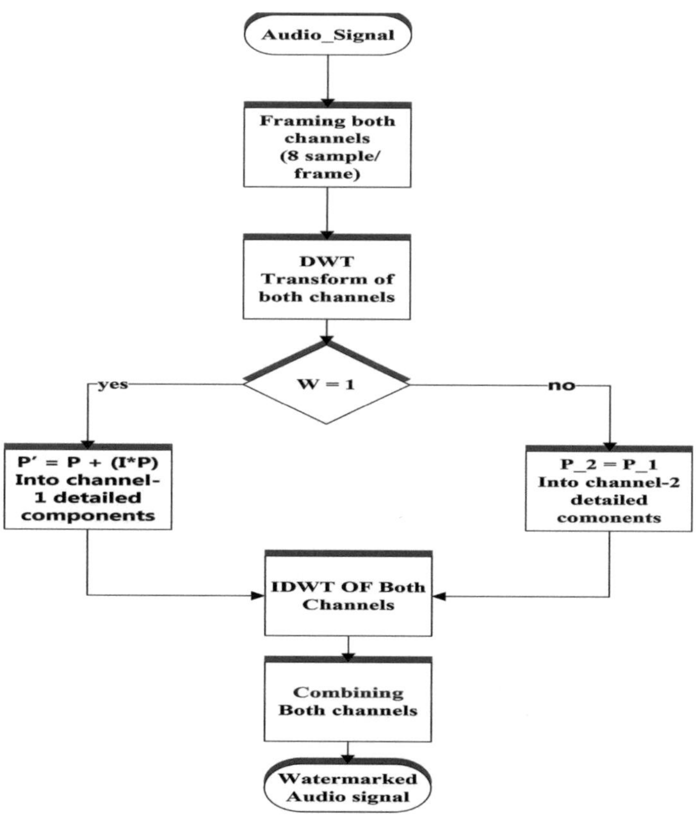

Figure 2.
Flowchart of the embedded process.

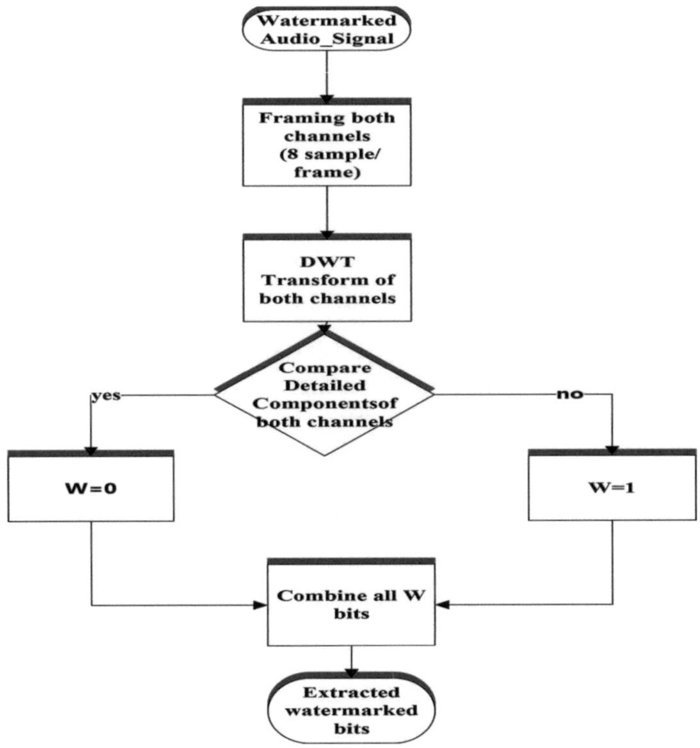

Figure 3.
Flowchart of the extraction process.

The flow chart of the embedded process is defined in **Figure 2** of the audio watermarking.

Step 4: After the completion of the embedding process at both channels, the inverse DWT transform is applied in both channels to get watermarked audio signal.

3.2 Extraction process of DWT-based audio watermarking

Input: watermarked audio signal; output: watermark

Step 1: Total 16 samples of both channels of the watermarked audio signal as an input is collected with similar steps followed in embedding process.

Step 2: DWT transform is obtained an approximate and detailed components of the both channel.

Step 3: Now the detailed part of the both channels are observed if they are same then our watermarked bit is 0 otherwise it is 1. The flowchart of the extraction process of the audio watermarking is shown in **Figure 3**.

Step 4: All the watermarked bit into single output to obtain the watermark which was embedded into an audio signal.

4. Hardware implementation of proposed audio watermarking

The architecture of watermark embedding process is defined in **Figure 4**. The process comprises of mainly three modules: DWT module, watermark embedding module and inverse DWT module. Initially, the audio samples are stored in Block RAM1 for the processing. The original watermark is also applied to watermark

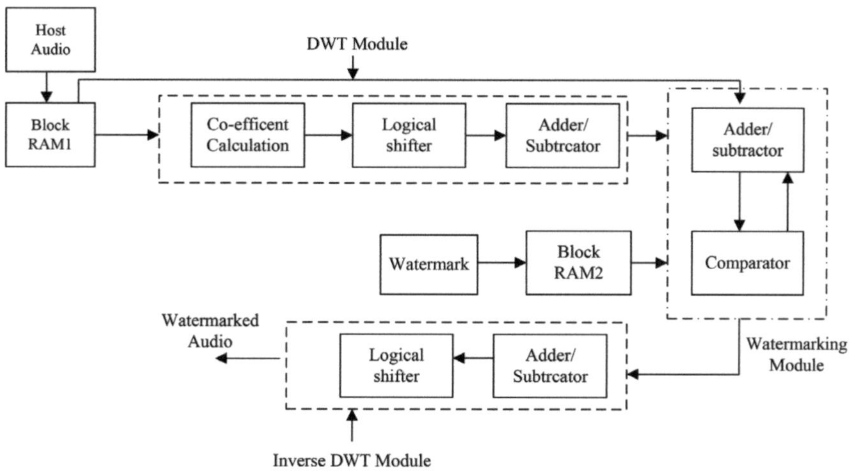

Figure 4.
VLSI architecture of watermark embedding process.

embedding unit. DWT module is used to read the values from the RAM and then converts these values to frequency domain co-efficient.

DWT module has coefficients calculation unit to compute various coefficient for Daubechies filter. After that, these coefficients are processed through watermarking module to insert the watermark. The watermarking module consists of comparator and adder/subtractor to embed the watermark into co-efficient. The comparator is used to take one bit of watermark from block RAM2 and as per bit, the module would decide to embed the watermark of the detailed co-efficient. After the insertion of watermark, the values are fed to inverse integer modules where watermarked audio samples are generated.

4.1 Hardware implementation of DWT

The models for executing of the DWT have mainly grouped in two classifications: (1) convolution filter based [15] and (2) the lifting based [15]. The vital discrete wavelet transform (DWT) is frequently refined by a convolution-based filter implementation using the FIR-filters for doing its transform [16]. FIR filters are applicable for improving the execution of the DWT hardware design [17]. Since a lifting structures have points of interest over a convolution-based regarding computation memory usage and complexity, more consideration is paid to the lifting-based approach. Daubechies and Sweldens [15] proposed the new wavelet scheme by taking into account of the second-generation wavelet. The lifting plan has better performance than convolution filter-based DWT. The lifting plan, which altogether depends on the spatial domain, has numerous favorable circumstances contrasted with filter bank structure, for example, lower complexity and power consumption with relatively reduced area. The lifting-based DWT has fundamental part of high-pass filter and divide the values in low-pass filters where sequence of upper and lower triangular matrices is being formed [18]. The lifting scheme contains mainly three stages, known as, split, predict (P), and update (U). Each of these steps is shown in **Figure 5**. The initial step is separating the original values into odd, and even samples, and then after the odd samples are changed to have the prediction and is obtained as the detail coefficients g_{j+1}. The even value is represented as coarser adaptation of input of the significant portion from determination. The average value of preserved signal, the detailed coefficients would help to revise the

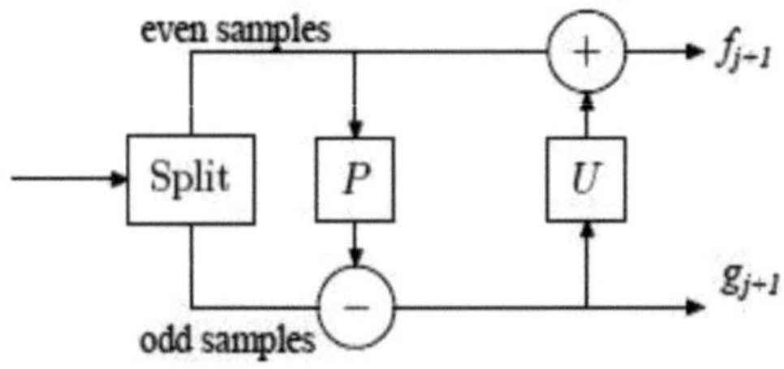

Figure 5.
Forward DWT process.

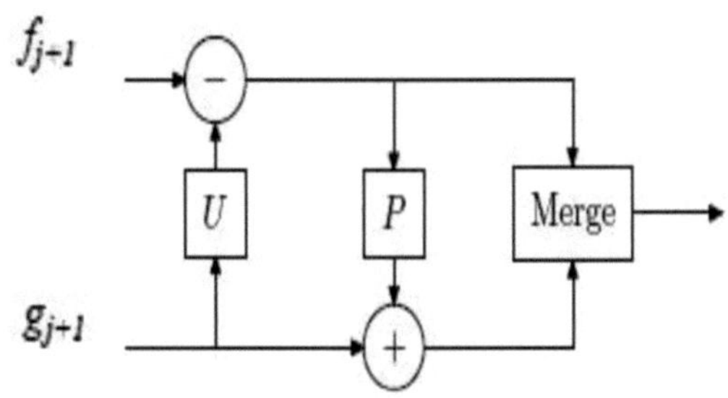

Figure 6.
Inverse DWT process.

even part. The process carried out in update step that creates f_{j+1} approximate coefficients. In order to achieve inverse transform, the sign is going to be exchanged at predict stage and the update stage and all operations are being applied in reversed order as defined in **Figure 6**.

The main objective is to achieve the lower and upper matrices (triangular type) and normalized diagonal matrix by dividing the polyphase matrix of the wavelet filters [19]. As indicated by the fundamental rule, the lifting filter of polyphase matrix of a 9/7 is defined as in Eq. (2).

$$P(z) = \begin{bmatrix} h_e(z) & g_e(z) \\ h_o(z) & g_o(z) \end{bmatrix}$$
$$(\lambda(z)\gamma(z)) = (x_e(z)z^{-1}x_o(z))P(z) \tag{2}$$

where g(z) and h(z) are high pass and low pass filter and is denied as notation of e (even) and o (odd) part respectively. The value is defined in Eq. (3).

$$P(z) = \begin{bmatrix} 1 & \alpha(1+z^{-1}) \\ 0 & 1 \end{bmatrix} \begin{bmatrix} 1 & 0 \\ \beta(1+z) & 1 \end{bmatrix} \begin{bmatrix} 1 & \gamma(1+z^{-1}) \\ 0 & 1 \end{bmatrix}$$
$$\begin{bmatrix} 1 & 0 \\ \delta(1+z) & 1 \end{bmatrix} \begin{bmatrix} K & 0 \\ 0 & 1/K \end{bmatrix} \tag{3}$$

where $\alpha(1+z^{-1})$ and $\gamma(1+z^{-1})$ are the predict polynomials, $\beta(1+z)$ and $\delta(1+z)$ are polynomials which are being updated, and scale normalization factor is denoted as K. And $\alpha = -1.586134342$, $\beta = -0.052980118$, $\gamma = 0.8829110762$, and $\delta = 0.4435068522$ are lifting co-efficient, and $K = 1.149604398$. For an input x(n) sequence, for $n = 0, 1, \ldots, N-1$, the steps of lifting scheme are given as in Eq. (4)

$$
\begin{array}{ll}
s_i^0 = x_{2n} & \bullet\ \text{Splitting into odd part} \\
d_i^0 = x_{2n+1} & \bullet\ \text{Splitting into even part} \\
d_i^1 = d_i^0 + \alpha\left(s_i^0 + s_{i+1}^0\right) & \bullet\ \text{Predict P1} \\
s_i^1 = s_i^0 + \beta\left(d_{i-1}^1 + d_i^1\right) & \bullet\ \text{Update U1} \\
d_i^2 = d_i^1 + \gamma\left(s_i^1 + s_{i+1}^1\right) & \bullet\ \text{Predict P2} \\
s_i^2 = s_i^1 + \delta\left(d_{i-1}^2 + d_i^2\right) & \bullet\ \text{Update U2} \\
d_i = \dfrac{d_i^2}{k} & \bullet\ \text{Scaling after then obtain detail component} \\
s_i = s_i^2 \times k. & \bullet\ \text{Scaling after then obtain approximate}
\end{array}
\qquad (4)
$$

Outputs d_i and s_i are low-pass as well as high-pass coefficients of wavelet.

4.1.1 Lifting scheme

Daubechies 9/7-based lifting scheme is shown in **Figure 7**. Each lifting step comprises one update as well as one predict step and that for second time for 2D implementation as P1, P2, and U1, U2, separately.

Pipelined shift-and-add logic plans multipliers used as a part of proposed DWT algorithm. This methodology diminishes the basic way essentially with little increment in latency and area. The shifter, signed adder and signed subtractor for multiplication process is used. For multiplication, alpha, beta, gamma, delta, multiply with K and divide with K module are discussed in **Figure 8**. The values are defined in Eq. (5)

$$
\begin{aligned}
\text{where, } |\alpha| &= 1 + \frac{1}{2} + \frac{1}{16} + \frac{1}{32} = 1.59375 \\
|\beta| &= \frac{1}{16} = 0.0625 \\
|\gamma| &= \frac{1}{2} + \frac{1}{4} + \frac{1}{8} + \frac{1}{128} = 0.8828125 \\
|\delta| &= \frac{1}{4} + \frac{1}{8} + \frac{1}{16} + \frac{1}{128} = 0.4453125 \\
|K| &= 1 + \frac{1}{8} + \frac{1}{16} = 1.875 \\
\left|\frac{1}{K}\right| &= \frac{1}{2} + \frac{1}{4} + \frac{1}{8} = 0.875
\end{aligned}
\qquad (5)
$$

where, $'S \gg n'$ defines the right shift to n bits, where $|\alpha| \times S = S + (S \gg 1) + (S \gg 4) + (S \gg 5)$. The predict step from first lifting generate odd and

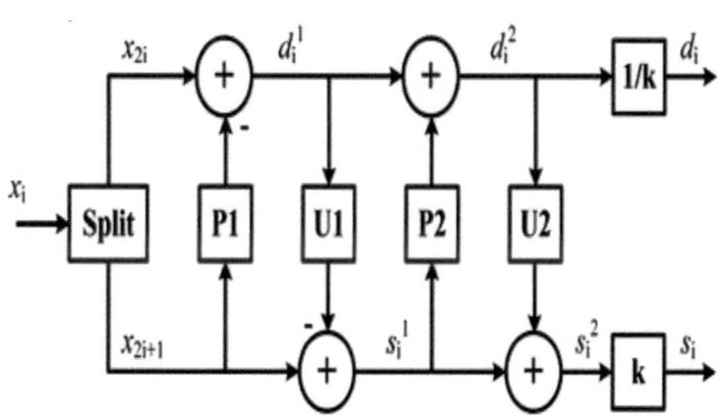

Figure 7.
Lifting plan for Daubechies 9/7 filter.

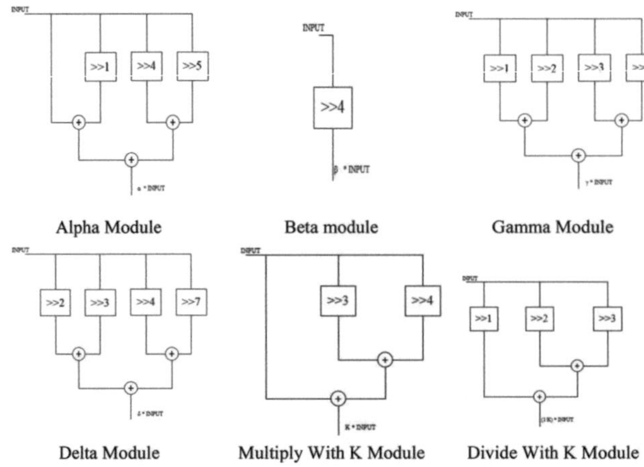

Figure 8.
DWT coefficients calculation.

even contribution after one clock cycle delay. The even value is included with past even input sample (s_i^0, s_{i+1}^0). Then operation of multiplier using primary filter coefficient is done at delay of the second and third clock cycles by applying only shifting and adding operation. After fourth clock cycle, the result of multiplier is considered at odd input sample (d_i^0) to update coefficient (d_i^1). At the end of fifth clock, the present value of predict (d_i^1) and the past value of the predict (d_{i-1}^1) with help of past even info (s_i^0) provides the first value of update (d_i^2). The adders is only the required operation at every clock cycle, thus critical path is defined through an adder delay only. The both phase, predict as well as update, of both stages are implemented in full pipelined approach to increase the speed. The overall lifting implementation comprises of four shifters and seven adders/subtractors. Moreover, the second stage of lifting have overall eight shifters and ten adders. The detail process is defined as in Eq. (6).

For an inverse DWT transform we use alpha, beta, gamma, delta module as same as discussed earlier but we use multiply module with the detailed coefficients and divide module with the approximate coefficients then we go reverse order of all the equation and finally we obtained original audio sample. Total eight samples for DWT transform are considered so after the inverse DWT transform eight samples are obtained. All the inverse DWT transform equation steps are discussed under.

$$\begin{aligned}
s_i^0 &= x_{2n} \\
d_i^0 &= x_{2n+1} \\
d_i^1 &= d_i^0 + \alpha\left(s_i^0 + s_{i+1}^0\right) \\
s_i^1 &= s_i^0 + \beta\left(d_{i-1}^1 + d_i^1\right) \\
d_i^2 &= d_i^1 + \gamma\left(s_i^1 + s_{i+1}^1\right) \\
s_i^2 &= s_i^1 + \delta\left(d_{i-1}^2 + d_i^2\right) \\
d_i &= \frac{d_i^2}{k} \\
s_i &= s_i^2 \times k.
\end{aligned} \tag{6}$$

The proposed design of DWT and inverse DWT would help to have efficient audio watermarking algorithm (**Figure 9**).

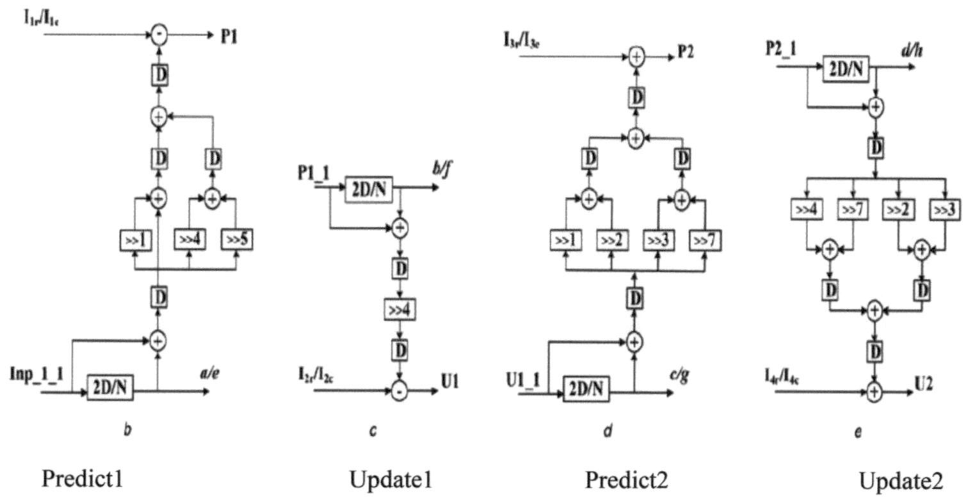

Figure 9.
Predict and update implementation of lifting scheme.

5. Simulation and result

The results are initially developed through MATLAB and then hardware implementation are achieved to verify the real-time implementation.

5.1 MATLAB

Experiments are performed in MATLAB 2010a. The proposed algorithm uses classical/pop music and speech audio clips in order to evaluate the performance [20]. These are three different types of audio clips are considered as they have different characteristics, perceptual properties and energy distribution. These audio signal have various distinct characteristics and also contains some selective features such as low energy, pulse clarity, pitch (in Hz.), inharmonicity, sampling rate (in Hz.), zero crossing rate (per second), spectral irregularity, temporal length (seconds/sample), tempo (in bpm), rms energy, etc. Each audio sample is of mono file of a 16-bit with sampling rate of 44.1 kHz of WAVE format. The watermark is of binary image of a 30 $\times$ 30 bits as in **Figure 10**. The synchronization code is a 16-bit Barker code of value "1111100110101110". The wavelet is applied with two decomposition levels. Array size m is 50 and the range of quantization step size Δ starts from 0.15 for speech audio and goes up to 0.6 for pop audio signal. The performance of audio watermarking algorithms is quantified by robustness, payload and inaudibility parameter [21].

Figure 10.
Watermark.

The inaudibility is measured using signal to noise ratio (SNR). It is a used to calculate the similarity between distorted watermarked audio signal and undistorted original audio signal. SNR is calculated as in Eq. (10):

$$SNR = -10log_{10}\left[\frac{\Sigma_i f_i^2}{(\Sigma_i(f_i' - f_i)^2)}\right] \tag{7}$$

where f_i is original audio signal, whereas f_i' is watermarked audio signal. It helps to calculate the noise induced in the watermark and defines the inaudibility.

Robustness: normalized correlation (NC) measure the similarity between original and extracted is given by:

$$NC(w, w') = \frac{\sum_{i=1}^{M}\sum_{j=1}^{M} w(i,j)w'(i,j)}{\sqrt{\sum_{i=1}^{M}\sum_{j=1}^{M} w^2(i,j)}\sqrt{\sum_{i=1}^{M}\sum_{j=1}^{M} w'^2(i,j)}} \tag{8}$$

here w is original watermark, w′ defines the extracted watermark, and i and j are indices to represent the watermark image. Generally, NC is to be considered as equal to 1. The robustness performance is measured using bit error rate (BER) as in Eq. (9).

$$BER(w, w') = \frac{\text{Number of error bits}}{\text{Number of total bits}} \times 100\% \tag{9}$$

The different attacks are considered for the robustness measurement of our proposed algorithm. The detailed of each signal processing attacks are defined and results are defined in **Table 1** [22].

a. **Re-quantization**: original watermarked audio signal of 16 bit/sample is down re-quantized at 8 bits/sample, which further back quantized to 16 bits/sample.

	Pop		Speech		Classical	
Attack	**NC**	**BER**	**NC**	**BER**	**NC**	**BER**
No Attack	1	0	1	0	1	0
Re-quantization	1	0	1	0	1	0
AWGN	0.999	0.003	0.995	0.005	0.999	0.002
Low-pass filter	0.999	0.001	0.997	0.004	0.999	0.001
Re-sampling	1	0	1	0	1	0
Mp3 64 kbps	0.9821	0.041	0.9878	0.037	1	0
MP3 128 kbps	1	0	1	0	1	0
Random cropping	0.997	0.002	0.999	0.001	0.998	0.002
Invert	1	0	1	0	1	0
Echo addition	0.997	0.002	0.998	0.003	0.999	0.002
Denoising	0.996	0.001	0.994	0.005	0.996	0.001
Pitch shifting	0.999	0.001	1	0	1	0

Table 1.
Experimental results for robustness of proposed algorithm.

b. **Additive white Gaussian noise (AWGN)**: to evaluate performance, Gaussian noise is inserted in the watermarked signal till an SNR reaches to 20 db.

c. **Low-pass filtering**: Butterworth filter of second-order is used at 11,025 Hz cutoff frequency.

d. **Re-sampling**: the sampling rate of the watermark signal is 44.1 kHz, further it is re-sampled at 22.05 kHz, and again sampled back at 44.1 kHz.

e. **MP3 compression 64 kbps**: the layer-3 compression of MPEG-1 is being applied. The audio signal with watermark is compressed with 64 kbps bit-rate and subsequently decompressed in the WAVE format.

f. **MP3 compression 128 kbps**: the layer-3 compression of MPEG-1 is being applied. The audio signal with watermark is compressed with 128 kbps bit-rate and subsequently decompressed in the WAVE format.

g. **Random cropping**: total sample of around 10% are cropped at randomly selected positions (front, middle and back).

h. **Invert**: all sample values are inverted in time domain with phase shift 180°.

i. **Echo addition**: an echo signal is added (with a decay of 41% and a delay of 98 ms) inside the watermarked audio signal.

j. **Denoising**: the audio signal with watermark is denoised with function of "automatic click remover" available in Adobe Audition 3.0.

k. **Pitch shifting**: it is most difficult attack for audio watermarking algorithms, because it tends to shift frequency fluctuation. In the results, the pitch is being shifted around one higher degree and one lower degree. These are applied to all three audio signals are shown in given in **Table 1**.

The payload data of the proposed algorithm is shown as:

$$B = {}^{R}/_{m \times 2^{k}} \text{bps} \tag{10}$$

The data payload is considered as 220 bps.

5.2 Comparison with related work

The general comparison is made between our proposed method and two similar methods [23] and is given in **Table 2**. As per reported results in **Table 2**, our proposed algorithm has higher capacity of embedding and lower value of BER. The proposed algorithm may achieve higher performance by reducing payload (which would be achieved by decomposition level increase for wavelet transform or length of array increases). The strength for embedding would increase with that approach.

5.3 Hardware results

The architecture is designed and implemented using Verilog HDL and targeting vertex 5 xc5vlx20t-2ff323 FPGA. We synthesized on Xilinx ISE 14.7. Each input and

Algorithm	Proposed	[24]	[25]
Payload	220	172	196
Noise reduction	0	0	0
BER	20 dB	20 dB	20 dB
Cropping and shifting (robust)	Yes	Yes	No
MP3 64 kbps (BER)	0.041	0.0434	0.01

Table 2.
Comparison with related audio watermarking.

Resources	DWT	Watermark embedding	Inverse DWT
BELs	13,214	5026	10,586
Registers	4946	1258	4268
Adders/subtractor	507	130	468
Multiplier	0	0	0

Table 3.
FPGA report for device utilization.

Resources	Total	DWT	Watermark Embedding	Inverse DWT
Slice	16840	1980 (11%)	570 (3%)	1796 (10%)
Slice FFs	33280	498 (1%)	137 (1%)	412 (1%)
4 input LUTs	33280	3782 (11%)	956 (2%)	3584 (10%)
Bounded IOBs	519	158 (30%)	52 (10%)	135 (26%)
GCLKS	24	1 (4%)	1(4%)	1 (4%)

Table 4.
FPGA report for resource utilization.

Parameter	DWT	Watermark embedding	Inverse DWT
Maximum frequency	36.12 MHz	47.26 MHz	39.32 MHz
Minimum period	27.68 ns	21.15 ns	24.32 ns

Table 5.
Synthesis report for timing analysis.

output is defined using IEEE 754 SP format because of complex calculations during DWT and inverse DWT. **Tables 3** and **4** provides the hardware utilization of targeted FPGA, and **Table 5** has the total computation delay. During synthesis process, the proposed audio watermarking scheme is validated for the real-time performance by FPGA prototyping.

6. Conclusion

The audio watermarking algorithm is proposed for different audio application. The algorithm uses DWT transform during watermarking process. The proposed algorithm has blind detection and has admirable performance against the attacks. A discrete wavelet transform (DWT) represents a data points regarding wavelet at various frequencies. In this chapter, hardware architecture of DWT-based digital audio watermarking is also developed which is used for real-time applications. Digital audio watermarking used in many applications like ownership protections, Tamper detection and localization, and media forensics. Above analysis shows that proposed algorithm gives good SNR with higher inaudibility and NC is also almost equal to 1. Various attacks are considered to check the robustness of the algorithm. This algorithm produces excellent resistance to many attacks. The FPGA prototyping is done for hardware performance measurement for real-time application.

Author details

Amit M. Joshi
National Institute of Technology, Jaipur, Rajasthan, India

*Address all correspondence to: amjoshi.ece@mnit.ac.in

References

[1] Joshi A, Mishra V, Patrikar RM. Real time implementation of digital watermarking algorithm for image and video application. In: Watermarking—Volume 2. Rijeka, Croatia: IntechOpen; 2012

[2] Xiang-yang W, Pan-pana N, Ming-yu L. A robust digital audio watermarking scheme using wavelet moment invariance. The Journal of Systems and Software. 2011;**84**:1408-1421

[3] Hwang M-J, Lee JS, Lee MS, Kang H-G. SVD-based adaptive QIM watermarking on stereo audio signals. IEEE Transactions on Multimedia. 2018;**20**(1):45-54

[4] Joshi AM, Gupta S, Girdhar M, Agarwal P, Sarker R. Combined DWT—DCT-based video watermarking algorithm using Arnold transform technique. In: Proceedings of the International Conference on Data Engineering and Communication Technology; Singapore: Springer; 2017. pp. 455-463

[5] Joshi AM, Mishra V, Patrikar RM. Design of real-time video watermarking based on integer DCT for H. 264 encoder. International Journal of Electronics. 2015;**102**(1):141-155

[6] Hu H-T, Hsu L-Y. A DWT-based rational dither modulation scheme for effective blind audio watermarking. Circuits, Systems, and Signal Processing. 2016;**35**(2):553-572

[7] Hu H-T, Hsu L-Y. Incorporating spectral shaping filtering into DWT-based vector modulation to improve blind audio watermarking. Wireless Personal Communications. 2017;**94**(2):221-240

[8] Li J-f, Wang H-X, Wu T, Sun X-m, Qian Q. Norm ratio-based audio watermarking scheme in DWT domain. Multimedia Tools and Applications. 2018;**77**(12):1-17

[9] Lalitha NV, Vara Prasad P, UmaMaheshwar Rao S. Performance analysis of DCT and DWT audio watermarking based on SVD. In: 2016 International Conference on Circuit, Power and Computing Technologies (ICCPCT); Nagercoil, India: IEEE; 2016. pp. 1-5

[10] Kotteri KA, Barua S, Bell AE, Carletta JE. A comparison of hardware implementations of the biorthogonal 9/7 DWT: Convolution versus lifting. IEEE Transactions on Circuits and Systems II: Express Briefs. 2005;**52**(5):256-260

[11] Algredo-Badillo I, Castillo-Soria FR, Ramirez-Gutierrez KA, Morales-Rosales L, Medina-Santiago A, Feregrino-Uribe C. Lightweight security hardware architecture using DWT and AES algorithms. IEICE Transactions on Information and Systems. 2018;**101**(11):2754-2761

[12] Goran S, Prokin M, Rajović V, Prokin D. Novel one-dimensional and two-dimensional forward discrete wavelet transform 5/3 filter architectures for efficient hardware implementation. Journal of Real-Time Image Processing. 2016:1-20

[13] Singh R, Mohanty SR, Kishor N, Thakur A. Real-time implementation of signal processing techniques for disturbances detection. IEEE Transactions on Industrial Electronics. 2019;**66**(5):3550-3560

[14] Dragoi I-C, Coltuc D. Adaptive pairing reversible watermarking. IEEE Transactions on Image Processing. 2016;**25**(5):2420-2422

[15] Kłos MJ. Determination of road traffic flow based on 3D Daubechies

wavelet transform of an image sequence. In: International Conference on Computer Vision and Graphics; Cham: Springer; 2016. pp. 573-580

[16] Tiwari VK, Jain SK. Hardware implementation of polyphase-decomposition-based wavelet filters for power system harmonics estimation. IEEE Transactions on Instrumentation and Measurement. 2016;**65**(7):1585-1595

[17] Eminaga Y, Coskun A, Kale I. Hybrid IIR/FIR wavelet filter banks for ECG signal denoising. In: 2018 IEEE Biomedical Circuits and Systems Conference (BioCAS); IEEE; 2018. pp. 1-4

[18] Darji A, Agrawal S, Oza A, Sinha V, Verma A, Merchant SN, et al. Dual-scan parallel flipping architecture for a lifting-based 2-D discrete wavelet transform. IEEE Transactions on Circuits and Systems II: Express Briefs. 2014;**61**(6):433-437

[19] Darji A, Agrawal S, Oza A, Sinha V, Verma A, Merchant SN, et al. Multiplier-less pipeline architecture for lifting-based two-dimensional discrete wavelet transform. IET Computers and Digital Techniques. 2015;**9**(2):113-123

[20] Olanrewaju RF, Khalifa O. Digital audio watermarking: Techniques and applications. In: International Conference on Computer and Communication Engineering (ICCCE 2012); 3–5 July 2012; Kuala Lumpur, Malaysia; 2012

[21] Al-Haj A. An imperceptible and robust audio watermarking algorithm. EURASIP Journal on Audio, Speech, and Music Processing. 2014;**2014**:37

[22] Karthigaikumar P, Jarline Kirubavathy K, Baskaran K. FPGA-based audio watermarking—covert communication. Microelectronics Journal. 2011;**42**(5):778-784

[23] Ozer H, Sankur B, Memon N. An SVD-based audio watermarking technique. In: Seventh ACM Workshop on Multimedia and Security; 2005. pp. 51-56

[24] Wu S, Huang J, Huang D, Shi YQ. Efficiently self-synchronized audio watermarking for assured audio data transmission. IEEE Transactions on Broadcasting. 2005;**51**(1):69-76

[25] Bhat V, Sengupta I, Das A. A new audio watermarking scheme based on singular value decomposition and quantization. Circuits, Systems, and Signal Processing. 2011;**30**(5):915-927